understanding body movement

understanding body movement

AN ANNOTATED BIBLIOGRAPHY

MARTHA DAVIS

ARNO PRESS
A New York Times Company
New York   1972

131398

Copyright © 1972 by Martha Davis

LC# 73-37652
ISBN 0-405-00286-6

Manufactured in the United States of America

CONTENTS

INTRODUCTION TO THE BIBLIOGRAPHY

Purpose, Scope, and Criteria for Title Selection

One hundred years after Darwin wrote his brilliant essay on expression of emotion (cf. ref. 188), there is a renaissance of research in body movement. Interest in expressive movement and nonverbal communication has increased so much in the last decade that these might be called the years of a "movement movement."

Research on the subject is growing within a number of disciplines, and there is a need for a bibliography of the literature on movement behavior which would facilitate this research and provide historical perspective. Most of the work in this field is done in relative isolation; there is little continuity of research and few "schools" of study. Consequently, even today a writer may speak of his research as if it were totally original and may regard the field as unexplored. It is important to know what has been done—and there is a surprising wealth of writing on the subject—to benefit from past efforts and to give credit where it is due. Hopefully this bibliography will contribute to the study of movement behavior in this way. It is a source book for students, behavioral researchers, and psychotherapists, as well as for performing artists and educators.

The term "movement behavior" is used throughout this book to refer to the anthropology and psychology of physical body movement. Books and articles on what is variously called expressive movement, kinesics, nonverbal communication, body language, and psychological aspects of movement, motor activity, and motility are cited.

The bibliography does not include titles on the neurology or physiology of movement, kinesiology, sensorimotor coordination, motor skill, or motor learning per se, unless these are explicitly related to psychological issues. Although motor learning and performance are, of course, a part of the psychology of movement, only a few related works are cited, either because they have good reviews of the literature in this area or because they deal with developmental, personality, or population patterns in motor performance or motor learning. Certain other areas related to movement, such as proxemics, body image, and dance ethnology, also are not dealt with extensively. However, care has been taken to cite representative literature and good reference lists on subjects closely related to movement behavior that are not included here.

The bibliography contains original abstracts of published and unpublished literature on body movement style, facial expression, gaze behavior, symbolic actions, gestures, postures, movement interaction, and psychological aspects of coordination, motor development, and abnormal movement. In most cases, the movement patterns discussed are visible at least through slow-motion film; however, EMG studies and experiments using special apparatus to measure motor activity are included. Because this is a book designed to aid research in movement behavior, special attention has been given to literature on movement notation and observation systems, structured motor tests, scales and inventories, and experimental equipment that may prove valuable in areas beyond those for which it was originally developed.

Priority has been given to books and papers that present data on or explicitly describe body movement patterns in some way. Papers were not selected on the basis of any judgment of the quality of the research itself. The main criteria for selection were that a work be in English, that it explicitly describe or refer to movement behavior, and that it relate in some way to psychological or anthropological aspects of body movement.

Included in the bibliography are some very early historical references, but the majority of the titles range from 1900 to June of 1971. A few works in press are cited. In most cases the references given are available in university or medical libraries. In this field, however, even many of the best works are rare, unpublished, or out of print. If a work cited is unpublished, an address where the reader may obtain copies is supplied. The "Datrix order numbers" of doctoral dissertations available from University Microfilms (Xerox Corporation, Ann Arbor, Michigan 48106) are listed. Beyond this, the decision has been made to list rare, hard-to-obtain works in the hope that they will be reprinted or otherwise made more available.

The annotations were done from library works available to the author. Frequently this meant that the books which were read were not the latest editions. When possible, the latest editions of books *which are in print* are cited in brackets in addition to the sources which the author used. Also for those books which are in print, the publishers' names and locations have been brought up to date.

This bibliography contains over 900 titles, annotated and subject-indexed. There is no way of knowing how complete the list is, but perhaps its scope can be illustrated with a description of how the titles were collected. One cannot go to the index of a reference book and look under such entries as "expressive movement," "kinesics," or "nonverbal communication" and find many titles. Such subject listings are rarely found in any relevant book. Searching for titles in reference books such as *Psychological Abstracts* yields surprisingly few works. (This is not true, of course, with regard to motor learning or motor performance.)

To collect the titles included in this bibliography, therefore, I began with private reference lists of a few researchers in the field, some special sources such as a computer search of doctoral dissertations (the Datrix service of University Microfilms), and a few large existing bibliographies.[1] Then the references in the books and papers themselves were scanned for titles. The collection grew to some 1,500 titles, their sources encompassing child development, psychology and psychiatry, ethology and animal behavior, anthropology, sociology, and the performing arts. Literature concerning dance, drama, and mime was considered particularly important, because these arts have accumulated knowledge about movement behavior for centuries. New areas such as dance therapy, body-awareness techniques, and special body movement training were also surveyed for works containing analyses and psychological interpretation of movement, posture, muscular tension, and so on.

From the collection of 1,500 titles, the present list of some 900 works was then selected according to the criteria mentioned above. In every case except the dissertations,[2] the original sources were read. While I had assistance in collecting the titles, checking source information, and preparing the manuscript, the selection, annotations, and

[1] Examples of these include: Yvonne Brackbill, ed., *Research in Infant Behavior: A Cross-Indexed Bibliography* (Baltimore: The Williams & Wilkins Company, 1964); Gordon W. Allport and Philip E. Vernon, *Studies in Expressive Movement* (New York: The Macmillan Company, 1933), pp. 249-260.

[2] Annotations for the doctoral dissertations were made from information in *Dissertation Abstracts*.

indexing are my own. If the bibliography has any inaccuracies or serious omissions, they are not intentional.

Acknowledgments

I am grateful to Ray L. Birdwhistell, Flora Davis, Adam Kendon, Albert E. Scheflen and Mark Zalk for making their reference lists available to me, with special thanks to Adam Kendon for his continual help and advice. Throughout the year that this bibliography was in preparation, I received suggestions of titles from a number of people, which I greatly appreciate. This work would have been impossible without the help of Jody Zaccharias in collecting titles and Jane Liff and Janet Lyon in preparing the manuscript. My gratitude also to Dori Lewis, Vice-President at Arno Press, for her vote of confidence and her editorial guidance.

And, most of all, thanks to my husband, Sergio Rothstein. This book is dedicated to him.

A HISTORICAL OVERVIEW

A survey of the literature on movement behavior may help to integrate what is presented piecemeal in the bibliography. It will also define the range of works reviewed. This introductory survey will concentrate on behavioral research of movement from Darwin's work in 1872 to current literature on nonverbal communication. The historical and performing arts references cited in the bibliography will not be included in this brief general history.

Before Darwin's time, one must go to historical descriptions of dances and mime, to books on manners and oratory, or to rare bits of literature for significant passages on movement behavior. The Romans and Greeks showed a great deal of interest in mime and the gestures of oratory. (180)* One could also examine old books on deportment or how to gesture in public speaking for information on the sociology of body motion. (139) There are a number of works on the dances of various periods which provide a glimpse into the social mores of their eras. (25) And there have been a few attempts to reconstruct rituals and body movements from ancient literature and art. (544) Some of these historical sources are cited in the bibliography, along with works by twentieth-century artists.

The understanding of movement in the performing arts has a history of its own. One could argue that the knowledge which dancers,

*The parenthetical numbers indicate references in the bibliography which are *representative* of the subject described, not necessarily the only or the most important work on the subject.

1

actors, and mimes have accumulated about movement has had no influence on the behavioral research cited in the bibliography—with three notable exceptions. In the area of dance ethnology, dancers, anthropologists, and, occasionally, dance notators have collaborated to a degree. (543) In the new field of dance (or movement) therapy, many former dancers are becoming mental health professionals and are beginning to participate in behavioral research. (493) In rare instances, the theories and movement analysis systems of a noted dancer or choreographer such as Rudolf Laban have influenced behavioral research. (550) However, the resources of the artist—his skill in observing, analyzing, and interpreting body movement—are rarely utilized in behavioral research. Suffice it to say that some of the "false starts" or limitations of movement research might have been avoided by consulting those whose medium is movement, rather than by adapting terms and methods from other disciplines.

Body Movement Research—1872 to the Present

The scientific study of movement behavior has its immediate roots in the work of physiognomists who attempted to read character from the structure of the head and physiologists who extended the study of anatomy to an investigation of the face muscles involved in expression. (64) But it was Darwin's *Expression of the Emotions in Man and Animals* (1872) which established facial and body movement patterns as a subject of serious scientific interest. (188) No work since Darwin's treatise has explored expressive movement with as much originality and breadth. Darwin gathered observations of facial expression and body movement from diverse sources such as anthropological field data, animal studies, and observation of mental patients., He postulated theories concerning the origins and functions of movement patterns and expressions in man and animals, and some of the research problems he defined have been unexplored until recently.

The list of those who have written about expressive movement or nonverbal communication since 1872 reads like a "Who's Who" in the behavioral sciences; yet writers still defend the relevance of such study or introduce the subject as if it were esoteric and unheard of. It is as if a great many serious behavioral scientists have shown a fleeting interest in body movement and then gone on.

After Darwin the field of movement research, as it is considered here, developed into several independent areas. Most of the behavioral

literature cited falls into one of eight areas, which are only now becoming interrelated and which may be designated as follows:

DEVELOPMENTAL PATTERNS

MUSCLE TENSION

EXPRESSION OF THE EMOTIONS

PERSONALITY AND PSYCHODIAGNOSIS

PSYCHOLOGICAL INTERPRETATION OF GESTURE

INTERACTION AND COMMUNICATION

CULTURAL CHARACTERISTICS

ANIMAL BEHAVIOR

The following is a survey of issues explored in each of these areas.

DEVELOPMENTAL PATTERNS

Detailed observation of the development of motor reflexes, locomotion and visual-prehensile coordination flourished in the 1930's, particularly through the work of Gesell and his colleagues. (14, 363) The
emphasis was on stages of normal development—when the child focused, smiled, sat up, walked alone, and so on—and many of the observations were incorporated into general developmental scales. Of
course, they were concerned with developmental processes in general,
not in body movement per se; but their work is of particular interest
here because they were among the first to use film for minute examination of movement. While the controversy about the relationship between I.Q. and level of motor ability persists, interest in defining developmental characteristics of movement waned after the 1940's.

The study of individual differences in children's motor patterns was
pioneered by Fries in her work on the "congenital activity type." (335)
Observation of individual differences in motor patterns can be found
at least as a passing reference in many motor studies of children and
adults, but it has received more extended, serious attention from Fries
and a number of researchers involved in the study of neonate response differences. (89) A third trend in developmental literature can
be found in psychoanalytic writing on stages of motor development
in relation to psychosexual development and character formation.
(672)

Experimental research, semistructured testing, and naturalistic observation of the movement of infants and children have been extensive,
thanks to the infant's lack of language facility. The movement of
neonates has been carefully observed in hospitals; the activity of infants and nursery children has been documented in detail; and in the

past two decades, the nonverbal interaction of mother and infant, particularly in the feeding situation, has received serious attention. (580) Notwithstanding studies of abnormal movement patterns such as tics or severe hyperactivity in children, most of the developmental research of their movement has dealt with normal maturation.

MUSCLE TENSION

Muscle tension has been defined and measured in many ways. It is a research subject that ranges from analysis of postural tensions in relation to personality, to electromyographic recordings of muscle activity during mental tasks. Patterns of muscle tension have been correlated with emotion, anxiety, mental effort, motivation, and personality characteristics. (242, 374) Luria stimulated a great deal of research on muscle tension with his studies of people under severe emotional stress. (599) Jacobson initiated sophisticated analysis of muscle tension changes in relation to thought processes and affect. (462) Muscle tension research has developed steadily over some forty years, and with it research apparatus to measure muscle activity mechanically or electrically. Although it is an intraorganismic and physiological measure of movement, simultaneous EMG recordings of both subject and interviewer attest to the interpersonal significance of fine changes in muscle activity (617). Today, using everything from highly trained observers to miniaturized telemetric equipment from space research, the study of muscle tension is complex and promising. (447, 510)

EXPRESSION OF THE EMOTIONS

Facial expression and body movement have a prominent place in literature on the nature of emotion and the recognition or judgment of affect. In the 1920's and 1930's many American and European psychologists were interested in the study of overt expression of affect and attitude, particularly whether someone could interpret another's feelings from his facial expression. (469) In effect, most of this early research on recognition of facial expression was an investigation of the ability of naïve observers to judge emotion and the degree to which they agreed. Subsequently research was conducted on the ability to make facial expressions, what expressions actually occur in emotional situations, developmental stages in facial recognition and expressiveness, the characteristics of the judges, and the judgment of body postures. The nature-nurture controversy emerges clearly in this research. Studies showing that blind-deaf children have normal facial expressions were countered by studies indicating facial expressions were learned and that observers couldn't agree unless they knew the context in which the expression occurred. (376, 559) After World War II, study of facial expression reappeared in diverse areas. With more sophisticated statistical analyses, there were a number of attempts to determine primary dimensions or factors of facial expression (and, by extension, dimensions of emotion). (771) By the 1970's recognition of facial expression was being studied cross-culturally; ethologists were recording facial expressions in primates; and analysis of facial movement had become an integral part of kinesics research on communication patterns. (18, 86, 271) Recognition studies are now often conceived of in terms of the nature of social perception and the characteristics of the perceiver; and facial expression itself is studied increasingly as communication signals or cues. (836)

Parallel to, and often interrelated with, research on recognition of facial expression has been the literature on the nature of emotion. Numerous theoretical formulations about emotion have considered the role of motor processes—muscle tension changes, facial expressions, and body postures—in emotional situations. (126, 563)

PERSONALITY AND PSYCHODIAGNOSIS

Early psychiatrists considered abnormal movement patterns such as grimaces, catatonic rigidity, mannerisms, and stereotypies symptomatic of schizophrenia. (100) The correlation of abnormal movement patterns with psychotic or severely neurotic syndromes was carried on by psychoanalysts in papers on tics, compulsive rituals, and hysterical reactions, and in rare research studies such as Charlotte Wolff's. (305, 909)

However, the more general question of a relation between movement patterns and personality has stimulated surprisingly little research. Although it is an age-old assumption that how one moves, walks, gestures, etc., reflects his personality, serious study of this correspondence has been limited. In *Studies in Expressive Movement*, Allport and Vernon investigated whether individuals have characteristic movement styles as reflected in how they perform a series of mo-

tor and graphic tasks. (8) Their results were promising; yet few researchers have since expanded upon their research. "Psychomotor personality tests" that have been devised vary greatly in form and sophis-
tication, but these remain relatively obscure research tools. (837) The
observation of individual differences in movement patterns is cited as
a passing reference in many experimental studies of movement; however, the most important work on movement and personality came
from psychoanalytic theory, not from research.

In his *Character-Analysis*, Wilhelm Reich proposed that one's character defenses, personality, and style of relating to others are manifested concretely and visibly in his muscular tensions, postures,
breathing patterns, and expressive style. (726) Reich analyzed how
breathing and muscle tensions play an important role in character
formation and development of defenses. His clinical observations
and analysis of muscular tensions and postures have been greatly expanded by his colleague Lowen and have been given further research
attention by Christiansen. (152, 595) A few other psychoanalysts
have studied the relationship between movement patterns and character formation largely from observation of children. (514) Recently
the comparative study of movement style and personality is being
developed through use of a complex system of movement analysis
borrowed from dance. (45)

PSYCHOLOGICAL INTERPRETATION OF GESTURE

When people are casually interested in expressive movement, they
may ask, "Well, what does that mean?" By the question they often
mean, what does a particular gesture indicate about one's personality or attitudes? In the 1930's a few psychologists studied everyday
gestures and "nervous habits" such as rubbing the face, finger-fidgeting,
or shifting in one's seat. (698) They tabulated types of gestures observed and compared them with psychological traits of the subjects.
Krout attempted experimental investigation of the psychological
meaning of specific gestures and mannerisms. (535)

Freud (330) was one of the first to point out that everyday "chance
acts" have psychological significance; but psychoanalytic study of
gestures and positions did not develop until Deutsch's formulation of
"analytic posturology" in the late 1940's. (212) He observed patients'
movements and positions in psychoanalysis and correlated them with
what the patient was saying at the time as well as with aspects of his

history and personality. Deutsch proposed that certain positions and gestures reveal specific intrapsychic conflicts, sexual problems, attitudes, and so on. In the 1960's, using films of psychotherapy sessions, a number of researchers continued investigating the "meaning" of specific gestures and correlated them with speech content or with psychological factors assessed independently of the movement analysis. (608)

INTERACTION AND COMMUNICATION

As has been said, body movement has received increasing research attention in the 1960's and 1970's. But this development was not initiated from within the areas of psychology or child development, and is not primarily related to the "traditional" topics discussed: expression of emotion, character formation, and psychological meaning. During the 1960's, there occurred a shift in terminology and emphasis from "expressive movement" to "nonverbal communication," from the expressive-intrapsychic-individual aspects of movement behavior to its role in group interaction and communication. The prime mover in this transition is the anthropologist Ray L. Birdwhistell, who coined the term "kinesics" for such research. (78) Drawing upon structural linguistics and cultural anthropology, Birdwhistell argues that body movement is one "channel" of communication, consisting of culturally learned units which are patterned in ways analogous to language and which serve in the maintenance, regulation, and definition of face-to-face group interaction. The "meaning" of one's movement is not to be found in individual psychodynamics or in concepts of emotion, attitude, or psychophysical processes, but at the group level in the effects the behavior has and its place in the stream of communication. The kinesics research of Birdwhistell and a number of psychiatrists interested in small-group and family communication has resulted in discoveries of patterns of synchrony, regulation, and organization of group behavior that are extraordinary in their ramifications. (166, 766)

By the late 1960's, studies of body movement and gaze behavior in interaction were growing rapidly in number: the intricate ways in which movement accompanies speech and regulates interaction (503); how gaze behavior relates to liking, dominance, anxiety, or dependency (291); how distance and eye contact interrelate (28); and how postures, facial expression, and visual behavior indicate status, responsiveness, and attitudes toward the one addressed (655). Although many doing such research may not support Birdwhistell's theory of a parallel between linguistic and kinesic organization, and may not maintain so explicitly that movement patterns are culturally learned, Birdwhistell's influence is apparent, particularly in countering earlier limitations in movement research (e.g., studying only the face or hands, disregarding visual behavior, focusing on the individual in vacuo).

CULTURAL PATTERNS

Although closely related to communication research, the study of cultural characteristics and differences in movement patterns is a topic in its own right. As with many other areas of research, it too has a historic work that remains unique and isolated, namely, Efron's study of the differences between Italian and Jewish gestures. (254) The greatest source of information about cultural differences in body movement, however, is found in dance ethnology and in the observations of anthropologists on the role of dance and ritual in culture. (103) The recent "choreometrics" research of Lomax, Bartenieff, and Paulay on cross-cultural styles of dance and work integreates this tradition with the perspectives of Birdwhistell and the approach to movement analysis of Rudolf Laban. (587)

Dance ethnology, anthropological research such as choreometrics, and, to an extent, kinesics study of subcultures focus on the movement characteristics and patterns of a culture and the differences between cultures. A contrasting focus is that of the investigation of pan-cultural patterns of expression, such as the research of Ekman et al. on pan-cultural recognition of facial expression and Eibl-Eibesfeldt's studies of similarities in human and primate greeting behavior. (257, 274)

ANIMAL BEHAVIOR

Like Darwin, modern ethologists have an eye for patterns of "facial displays" and body movements in animals which are more than reminiscent of those in humans. Given strong impetus by early primate studies and Lorenz's analysis of the motor patterns of lower mammals and birds, ethologists have in effect resumed Darwin's studies of animal expression. (220) From field studies of primates to a minute analysis of the bee dance, there is a rich and fascinating literature about animal behavior, including works that devote serious attention to body motion. Thus, for example, they report that monkeys and apes greet, smile, frown, attack and defend in symbolic actions, and have complex patterns of synchrony, regulation of distance, and social hierarchy observable within their movement interaction. (137, 220) Like those who study human infants, ethologists are not caught up in verbal behavior, and they describe animal expression and body communication signals with a freedom and matter-of-factness often lacking in studies of human movement.

To speak of a history of research in movement behavior is somewhat misleading because, beyond individuals working in relative isolation or limited schools of interest in a specific topic, the study of expressive movement and nonverbal communication is only now gaining wide recognition as a field. Some of the authors cited in this bibliography probably have never considered their work as body movement research, and the list of those who have written directly about the subject may be surprising. Today all manner of movement is analyzed—visual behavior, facial expression, hand gestures, postures, muscle tension, etc.—with some sense of its belonging under the rubric of body motion or nonverbal communication. Certainly, the existence of expressive movement and nonverbal communication has been acknowledged for centuries; but the identification of "movement analysts," "kinesicists," and "movement (or dance) therapists," those people whose research or therapeutic work involves an understanding of the psychology or anthropology of movement, is a phenomenon of the last decade.

Of course, the areas or topics delineated here largely parallel the traditional disciplines of physiological psychology, child development, general psychology, psychiatry, anthropology, and ethology. And the history surveyed in relation to movement behavior largely parallels the history of what has interested behavioral researchers in the United States and Europe in this century. This is especially clear in the shift from a focus on physiological aspects to an interest in affect, personality, and intrapsychic processes to a concentration on interpersonal and cultural-level phenomena. With the increased use of film, those interested in what "really" goes on in complex, face-to-face interaction—particularly group and family psychotherapists and anthropologists in the field—can now capture what occurred at a specific time and observe it over and over in depth. Obviously movement research has been facilitated by and has increased with the technological advances in photography, filmmaking, and videotape. And television itself has helped shift everyone's attention from words and sounds to pictures and motion.

But there must be deep historical and cultural reasons for the delay in the "movement movement." Why there is a renaissance of interest in facial expression and body movement now is an interesting and largely unexplored question in its own right. Observing movement has been considered an invasion of privacy; movement itself is largely ignored and relegated outside awareness—and these are reactions which need to be better understood and respected. Notably, some writers of the encounter group and human-potential movement have addressed themselves to this renewed interest in the body. (776) They rebuke the overintellectualized, goal-oriented, and highly verbal Western mentality that devalues the physical, the nonverbal, the expressive.

Whatever the historical reasons why it should come of age now, the field is clearly becoming extraordinarily interesting and diverse. There are already theoretical camps debating innate vs. learned, individual vs. cultural, expression vs. communication. The oversimplification in these debates is countered by the complexity of the subject. The problem is becoming one of determining which aspects of movement relate to which levels and dimensions of behavior, not whether body movement is the legitimate realm of any one discipline.

A bibliography of literature on body movement is really words about words about what is nonverbal; thus, there is a danger of losing sight of the original event. By its very nature, movement is elusive and hard to capture in words. One often has to visualize or imitate through his own movement the behavior a writer is discussing in order to be able to follow his analysis intelligently. It is not a field that grows primarily through its literature. What schools of movement research there are—from the facial recognition studies of the 1930's to the kinesics communication research of today—have depended primarily on photographs, films, videotapes, or direct collaboration. Most other work on the subject has been done by individuals in isolation.

If the field continues to grow as it has in recent years, films or videotapes will regularly accompany academic courses and books on

the subject. If this bibliography could be made visual in order to better convey the content of the literature, if the observations made by
those cited could be captured on film in a montage of movement patterns demonstrating the scope and humanity of the subject, it would
be a thousand times more compelling than all the words reviewed.
Such a film would be a silent stream of images and movement sped
up, slowed down, repeated to best convey what the writer has observed:

> How people around the world greet each other (257); the
> expression of a blind-deaf child as she discovers a doll
> (376); the visual behavior of dominant vs. dependent
> individuals (291); how a child reaches and grasps at
> four months, eight months, twelve months (401); a mime
> telling a story (34); how Hitler moved when he heard
> that the Germans had invaded France (673); the dance
> of Death in the *Green Table* ballet (796); how monkeys
> react to the facial expressions of other monkeys (665);
> what really happened as two· hitchhikers were passed
> up on the way to Reno (87); how women sit and how
> men sit (762); the bee dance indicating where the food
> is (343); how a mother moves synchronously with her
> healthy daughter and "cuts off" her schizophrenic daugh-
> ter (163); diverse cultural styles of working and danc-
> ing (587); how primates greet each other (467); . . .

BIBLIOGRAPHY

The letters *D*, *G*, *N*, *P*, *T* which appear throughout the Bibliography at the end of certain entries refer to drawings or diagrams, graphs, notation of movement, photographs, and tables, respectively.

1   **Abelson, Robert P., and Sermat, Vello.** "Multidimensional Scaling of Facial Expressions," *Journal of Experimental Psychology* 63 (1962):546-554. *T*

An experiment evaluating the efficacy of Schlosberg's three scales of facial expression—pleasant-unpleasant, attention-rejection, and tension-sleep—concluding that two dimensions account for most of the variance (pleasant-unpleasant and tension-sleep) and would be adequate in judgments of facial expression.

2   **Abraham, Karl.** "A Constitutional Basis of Locomotor Anxiety." In *Selected Papers of Karl Abraham, M.D.,* translated by D. Bryan and A. Strachey, pp. 235-243. London: The Hogarth Press, 1927.

Abraham proposes that patients who are phobic about walking alone, constitutionally have an "over-strong pleasure in movement" and motor rhythms which is not successfully repressed. Their inhibition of movement is related to incestuous fixations.

3   **Adams, F. R.** "Stereotype, Social Response and Arousal in a Case of Catatonia." *British Journal of Psychiatry* 113 (1967):1123-1128. *D*

"Motilograms," or oscillograph recordings obtained from a strain-gauge apparatus in the chair regularly used by a catatonic patient, showed the pattern of his constant rocking movements. The changes in amplitude and frequency of the rocking indicate that, although the patient appeared to be "staring into space and seemingly immersed in fantasy," he was in fact highly attentive to events around him and reacted differently to different individuals and events. The author discusses the function of high arousal in helping the patient maintain his mode of social withdrawal.

4   **Aldrich, Virgil C.** "Expression by Enactment," *Philosophy and Phenomenological Research* 16 (1955):188-200.

A theoretical paper on the significance of ritual and ceremony, with a critique of the theories of several "ritualists."

13

5 **Alexander, F. Matthias.** *The Use of the Self: Its Conscious Direction in Relation to Diagnosis, Functioning and the Control of Reaction.* New York: E. P. Dutton & Company, 1932. 143 pp.

Alexander describes the evolution of and rationale for his techniques for conscious correction of mental and physical habits such as body carriage or breathing. He analyzes "misdirection of use" and how one's actions become distorted by overemphasis on goals and faulty habits, and he illustrates his teaching and sensory awareness techniques in the treatment of a stutterer. A chapter on the importance of such principles for medical diagnosis and training is included.

6 **Alexander, F. Matthias.** *The Universal Constant in Living.* New York: E. P. Dutton & Company, 1941. 270 pp. *P*

Contains further exposition of the theory and practice of the Alexander technique.

The importance of one's "manner of use" and psychophysical activity and Alexander's modes of assessing and treating incorrect psychophysical habits of individuals with asthma, stuttering, tics, physical injuries, etc., are discussed. He also considers the relevance of his theories for physical education and preventive medicine. Much of this book is a review of and an answer to reactions to Alexander's work from various scientists.

7 **Allport, Floyd H.** *Social Psychology.* (Houghton Mifflin Company, 1924; reprint ed., New York: Johnson Reprint Corporation, 1967). 453 pp. *D G T*

Includes an analysis of the role of bodily expression and group formation in "contagion of emotion," a review of the gesture theory of the origin of language, and an excellent chapter on facial expression. In this last-mentioned chapter Allport assesses theories of Darwin, Wundt, and Piderit and reviews some studies of judgment of facial expression.

8 **Allport, Gordon W., and Vernon, Philip E.** *Studies in Expressive Movement.* New York: The Macmillan Company, 1933. 269 pp. *G T* [New York: Hafner Publishing Company, 1967]

A classic in this field and an exceptional example of controlled experimentation, direct objective measurement, and elaborate statistical analysis applied to the study of expressive movement. Allport and Vernon investigate whether there is a consistent motor style or integration in the expressive manner of the individual which reflects personality. Twenty-five male subjects, diverse in age and background, were given over thirty motor tests, twice each. The tasks included measures of tapping rate, muscle tension in the arm, average walking speed, space occupied by the subject's drawing, etc. While there was no evidence of a general motor factor, "group factors" derived from clusters of variables having the highest intercorrelation and internal consistency were found: namely, an "areal," or expansive, factor; a centrifugal, or "outward-tendency," factor; and an emphasis factor. Also of value are the theoretical discussions about problems of studying expressive style, the comprehensive classification of expressive movement terms, four case studies that clinically demonstrate consistency in movement and its correlation with personality, and a chapter on handwriting and personality.

9 **Altman, Stuart A.,** ed. *Social Communication Among Primates.* Chicago: University of Chicago Press, 1967. 392 pp. *D G P T*

There is often a great deal of information about facial expressions, postures, movements, and the contexts in which they occur scattered throughout books on animal behavior, and this one is a fascinating and beautifully illustrated example. A variety of behavioral scientists report on research done in the field, in captivity, or in the laboratory on primate social behavior, particularly sexual behavior, mother-child interaction, dominance patterns, group organization, and communication behavior.

10   **Ames, Louise Bates.** "The Constancy of Psycho-Motor Tempo in Individual Infants." *Journal of Genetic Psychology* 57 (1940):445-450. *T*

Films of eight infants, taken over three years, were analyzed to determine the tempo or speed with which the child manipulated different objects and climbed stairs or crept about. Using the number of film frames as the measurement, analysis showed great individual consistency over time and across activities. Also, speed of an action correlated with rate of development, and it did not increase appreciably with age.

11   **Ames, Louise Bates.** "Motor Correlates of Infant Crying." *Journal of Genetic Psychology* 59 (1941):239-247. *G P T*

A study of the movement characteristics of crying and non-crying behavior in a supine position in thirteen infants ranging from eight to forty-eight weeks old. Crying behavior was found to be characterized by "vigorous limb activity, unilateral arm and leg behavior, greater leg than arm activity, strong flexor tendencies . . ."; non-crying behavior by "limb extension, bilateral postures, greater arm than leg activity and holding of set postures" (p. 241). Some individual motor differences are described.

12   **Ames, Louise Bates.** "Supine Leg and Foot Postures in the Human Infant in the First Year of Life." *Journal of Genetic Psychology* 61 (1942):87-107. *D G P T*

Basing her work on a film study of normal infants, the author presents detailed descriptions of the characteristic positions and movements of legs, feet, and toes at each month of infanthood from four to fifty-two weeks. The analysis is largely in terms of flexion-extension-rotation patterns, reflexes, and laterality, with graphs and photographs illustrating what predominates at each interval, as well as a summary of the overall developmental trend.

13   **Ames, Louise Bates.** "Early Individual Differences in Visual and Motor Behavior Patterns: A Comparative Study of Two Normal Infants by the Method of Cinemanalysis." *Journal of Genetic Psychology* 65 (1944):219-226.   *G T*

The visual and motor behavior of two boys filmed throughout their first year of life and later at ages five and twelve was studied, and a summary of their different personality styles is presented. Patterns of visual regard and/or contact with objects and the rate of prehension (time from reach to grasp) and creeping were individually consistent over time.

14   **Ames, Louise Bates.** "Postural and Placement Orientations in Writing and Block Behavior: Developmental Trends from Infancy to Age Ten." *Journal of Genetic Psychology* 73 (1948):45-52. *P T*

An interesting analysis of the postures and arm positions observed in films of 179 children, ranging from thirty-six weeks to ten years old, while writing and using building blocks. The way the children performed these activities shows a clear developmental trend.

15   **Ames, Louise Bates.** "Development of Interpersonal Smiling Responses in the Preschool Years." *Journal of Genetic Psychology* 74 (1949):273-291. *D T*

Children from eighteen months to four years old were observed during nursery play behind a one-way vision screen, and each incidence of smiling or laughter and the context in which it occurred was noted. The kinds of body movements, activities, and social contexts which evoked smiling are noted for each age, together with the rate of smiling per age. Notably there is a change in which relationship most often evokes smiling: child in relation to his own gross motor activity, then child while approaching teacher, and at age four the child approaching other children. Individual differences in the amount of smiling are noted.

16   Ames, Louise Bates. "Bilaterality." *Journal of Genetic Psychology* 75 (1949): 45-50. *G T*

A summary of patterns of bilaterality and unilaterality in seven children from infancy to age ten. Repeated alternation of bilaterality and unilaterality can be seen at definite periods up to fifty-two weeks of age in both arms and legs and in prone and supine positions. Along with this, there is a progression from arms moving almost functionally as one to a differentiated use of each limb.

17   Anderson, J. D. "The Language of Gesture." *Folk-lore* 31 (1920):70-71.

A list of twenty-two common mudras, traditional gestures of the Hindu dance.

18   Andrew, Richard J. "The Origin and Evolution of the Calls and Facial Expressions of the Primates." *Behaviour* 20 (1963):1-109. *D P*

Proposing theories as to the causation and function of animal calls and facial expressions (e.g., that they may be protective responses evoked by marked stimulus contrast), Andrew describes various animal sounds and facial expressions and in what contexts they occur, primarily those associated with attack or defense, greeting, rage, exploration, and sexual activity. Although there is some information on expression in lower mammals and on infant-adult differences in humans, the major part of the article deals with calls and expressions observed in primates.

19   Andrew, Richard J. "Evolution of Facial Expression." *Science* 142 (1963): 1034-1041. *D P*

A succinct discussion of the causation and origin of facial displays in various mammals and primates; their function in informing other animals of one's imminent behavior or status; and their roots in protective responses, vigorous respiration, or grooming behavior. Photographs and drawings illustrate expressions and muscle actions involved.

20   Andrew, Richard J. "The Displays of the Primates." In *Evolutionary and Genetic Biology of Primates,* vol. 2, edited by J. Buettner-Janusch, pp. 227-309. New York: Academic Press, 1964. *D P*

Includes a review of literature on displays, evolutionary aspects of displays, and their relation to information transmission and social organization. Behaviors of protection, attack and submission, tail movements, marking and presentation, sexual activity, contact, grooming, and facial expressions are analyzed.

21   Andrew, Richard J. "The Origins of Facial Expressions." *Scientific American* 213 (Oct., 1965):88-94. *D P*

The author argues that facial expressions are not innate expressions of specific emotions but residues of functional actions, and he illustrates this theory with an analysis of the evolution of several expressions. For example, eyebrows raised in surprise is traced to ear flattening in mammals, which protects sensory organs, and scalp raising in primates, whose function is communicational. He also traces the evolution of the frown from lowering eyebrows in mammals, in order to focus on an object, to lowered eyebrows in primates and man, indicating a confident threat; the grin, from protective responses of biting and shrill crying to defensive grinning of monkeys and man in the presence of another who is dominant; and the smile, from the grin of the startle response which is unpleasant to a milder reaction to small, pleasurable degrees of stimulus change or surprise. He concludes with a discussion of "how human speech may have evolved from primitive facial and tongue movements."

22   **Andrews, Edward D.** *The Gift to Be Simple: Songs, Dances, and Rituals of the American Shakers.* (J. J. Augustin Publisher, 1940; reprint ed., New York: Dover Publications, 1962). 170 pp. *D*

The dances and rituals of the Shakers are described in detail, illustrated with diagrams and old prints, and interpreted regarding their history, meaning, and role in the Shaker culture.

23   **Andrews, Edward D.** "The Dance in Shaker Ritual," In *Chronicles of the American Dance,* edited by P. Magriel, pp. 2-14. New York: Henry Holt and Company, 1948. *D*

Includes a history of the Shakers from 1750 on, the sociology of their dances and ritual, and descriptions of the ritual dances and the variations that developed over time, with drawings illustrating them.

24   **Andrews, T. G.** "Some Psychological Apparatus: A Classified Bibliography." *Psychological Monographs,* Whole No. 289, 62 (1948):1-38.

A bibliography of 942 articles describing research equipment. Those for measuring body movement, muscle tension, and facial expression can easily be found through its index.

25   **Arbeau, Thoinot.** *Orchesography.* Translated by Mary Stewart Evans. New York: Dover Publications, 1967. 266 pp. *D*

A treatise first published in 1589, this beautifully reprinted edition is complete with illustrations, extensive editor's notes, and Labanotation and music scores of fourteen martial and social dances of the sixteenth century. Written as a dialogue between the Frenchman Arbeau and a young pupil he is instructing in the art of dancing, the book is a classic source on sixteenth-century dances and social mores.

26   **Ardrey, Robert.** *The Territorial Imperative: A Personal Inquiry into the Animal Origins of Property and Nations.* New York: Atheneum Publishers, 1966. 390 pp.

Unfortunately, in this important and very readable book, there are only scattered references to how animals move while defining and defending their territory. It is cited here primarily for its extensive bibliography of research on primate, mammal, and bird behavior, much of which addresses itself to body movement and facial expression.

27   **Argyle, Michael.** "Non-Verbal Communication in Human Social Interaction." In *Non-Verbal Communication,* edited by R. Hinde. London: Royal Society and Cambridge University Press, forthcoming. *D*

In this survey of the field of nonverbal communication (NVC), the author starts with discussion of current research approaches to NVC and the advantages of "rigorously designed experiments . . . in realistic settings" and recommended viewing arrangements and equipment. He then discusses literature on different kinds of nonverbal signals: bodily contact, proximity, orientation, appearance, smell, posture, head nods, facial expression, gestures, gaze, and tone of speech. A section on the function of NVC includes descriptions of what kinds of signals convey which attitudes, emotions, and self-presentations; how NVC sustains conversation;

and how personality, emotion, attitudes, and interaction are interpreted on the basis of NVC. NVC and the maintenance of interaction and differences between groups and individuals are discussed. The work concludes with a review of four different approaches to the origin of human NVC.

28   **Argyle, Michael, and Dean, Janet.** "Eye-Contact, Distance and Affiliation." *Sociometry* 28 (1965):289-304. *G T*

Following a list of previous findings on eye-contact (EC) behavior, the authors enumerate what they consider the main functions of EC. There are descriptions of two experiments supporting the theory that there is an equilibrium process between EC and distance: namely, length of glance increases with distance, and EC decreases with proximity. One experiment involved adult and child subjects standing "as close as is comfortable" while looking at a photo, a person with eyes shut and a person with eyes open. In a second study, dyads in conversation were seated 2 feet, 6 feet, or 10 feet apart, and EC was observed. There was less EC in mixed sex pairs. The article includes some discussion of body movement in relation to distance: for example, avoidance behavior such as covering the eyes occurs when two people are unusually close, and leaning forward occurs when they are sitting quite far apart.

29   **Argyle, Michael; Salter, Veronica; Nicholson, Hilary; Williams, Marylin; and Burgess, Philip.** "The Communication of Inferior and Superior Attitudes by Verbal and Non-Verbal Signals." *British Journal of Social and Clinical Psychology* 9 (1970):222-231. *G T*

Eighteen videotapes of two speakers making statements whose content reflected superior, equal, or inferior attitudes presented with nonverbal behavior intended to reflect one of the three attitudes were rated by observers. The verbal statement alone or a performer reading numbers in the three different manners had the same effect on shifting ratings. Also, women were found to be more responsive to nonverbal cues.

30   **Armitage, Merle.** *Martha Graham.* (Merle Armitage, 1937; reprint ed., Brooklyn, N.Y.: Dance Horizons, 1966). 132 pp. *D P*

Articles about Martha Graham and her dance form written in the 1920's and 1930's. Of note are the statements made by Miss Graham herself suggesting her aesthetic views and the way she interprets American movement styles.

31   **Armstrong, Edward A.** *Bird Display and Behaviour: An Introduction to the Study of Bird Psychology.* (Lindsay Drummond, 1942; reprint ed., New York: Dover Publications, 1965). 431 pp. *D P*

The display patterns or movements, postures, and sounds of various birds are described and analyzed for their social, emotional, and physiological significance in this prodigious and fascinating work. There are chapters on social ceremonies, social hierarchy, greeting, dominance, and territory behavior. Other subjects dealt with are "the psychological basis of nest-building," "the comparative study of display," "the function of emotion in behaviour," "the dances of birds and men," and "arena displays and the synchronisation of male and female rhythms."

32   **Arnheim, Rudolf.** "The Gestalt Theory of Expression." *Psychological Review* 56 (1949):156-171. *T*

The author reviews theories of expression and the development of expressive behavior and proposes the Gestalt view of the isomorphism and "structural similarity of psychical behavior." He notes briefly the ways that five dancers improvised on the words "sadness," "strength," and "night" to show how the dynamics of body movement may correspond to the dynamics of states of mind.

33   **Ascher, Edward.** "Motor Attitudes and Psychotherapy" *Psychosomatic Medicine* 11 (1949):228-234.

This presentation of seven case abstracts of psychiatric patients with various diagnoses focuses on the patients' motor symptoms and their interpretation.

34   **Aubert, Charles.** *The Art of Pantomime.* Translated by E. Sears. (1927; reprint ed., Bronx, N.Y.: Benjamin Blom, 1970). 210 pp. *D*

This fascinating book, which describes many variations in total body posture, hand and head gestures, and facial expression, illustrates these with good drawings and interprets their emotional and communicative meaning. There is also a section on "how to register the parts of speech" in movement. Clearly written, it exemplifies how much more advanced the performing arts can be in the study of movement behavior.

35   **Ayd, Frank J., Jr.** "A Survey of Drug-Induced Extrapyramidal Reactions." *Journal of the American Medical Association* 175 (1961):1054-1060. *G T*

Of 3,775 patients receiving phenothiazine tranquilizers, 38.9% developed extrapyramidal reactions such as akathisia, dyskinesia, and parkinsonism. There were marked differences in reaction according to sex and evidence that these patients were neurologically susceptible.

36   **Ayres, A. Jean.** "Patterns of Perceptual-Motor Dysfunction in Children: A Factor Analytic Study." *Perceptual and Motor Skills* 20 (1965):335-368. *T*

Following a review of literature on the relation between cognitive functioning and sensorimotor activity, the author reports on the results of an extensive battery of tests given to one hundred nonretarded children between six and eight years old who teachers reported had learning difficulties, hyperactivity or clumsiness, and lower performance than verbal scores on I.Q. tests, compared with a control group of fifty "normal" children matched for age and sex. Many of the thirty-five tests were drawing or perceptual tasks; however, there were also specific tests for eye and hand dominance, tendency to avoid crossing the body midline, kinesthetic perception, fine motor tasks graded in complexity, and tests of rhythm and degree of hyperactivity. Factor analyses of the dysfunction group data revealed specific patterns of dysfunction, including developmental apraxia (finger agnosia, deficiency in motor planning, etc.); perceptual dysfunction, including tactile and kinesthetic perception of position in space; tactile defensiveness (hyperactive behavior, etc.); deficient integration of function of the two sides of the body (difficulty crossing the midline, ability to discriminate the right and left sides, etc.); and visual figure-ground discrimination (with evidence that this is linked with motor processes as well as perceptual ones).

37   **Bailey, Flora L.** "Navaho Motor Habits." *American Anthropologist* 44 (1942): 210-234.

An analysis, based on live and film observation, of the ways in which Navaho Indians move. Generalizations about the body movement patterns and manner of Navahos in three areas of activity—personal habits (eating, sleeping, sitting, etc.), social habits (handshake, storytelling, etc.), and work habits (shearing sheep, spinning wool, etc.)—are based on observing the pattern in a minimum of four cases. The descriptions, detailed and often fascinating, usually include who does what, in what context, and in what way (what body parts are used, how an instrument is held, in what direction and with what force or tempo a movement is made). Among the other notes included are: how games are played, movement patterns in ceremonies, how motor habits are transmitted, notable restrictions of actions, and handedness.

38   **Bakan, Paul.** "The Eyes Have It." *Psychology Today* 4 (Apr., 1971):64-67. *T*

Research by Bakan, Merle E. Day, and others shows that individuals have a tendency to look either to the left or to the right when answering reflective questions. Bakan reports relationships between CLEM tendencies (conjugate-lateral eye movement) and personality (e.g., "left-movers" are more subjective, hypnosis-susceptible, emotional), sex differences, brain waves, levels of consciousness, and brain hemisphere dominance.

39   **Balint, Michael.** "Individual Differences of Behavior in Early Infancy and an Objective Method for Recording Them: I, Approach and the Method of Recording." *Journal of Genetic Psychology* 73 (1948):57-79. *T*

The evolution of a decision to study early individual differences through an objective method of analyzing sucking and feeding behavior is described. One hundred hospitalized and bottle-fed infants of different ages and with various illnesses were studied. The development of an apparatus inserted into the bottle which did not bother the baby and which was connected to a polygraph in order to record the patterns of pressure is elaborately described, along with notes on the various kinds of sucking patterns found.

40   **Balint, Michael.** "Individual Differences of Behavior in Early Infancy and an Objective Method for Recording Them: II, Results and Conclusions." *Journal of Genetic Psychology* 73 (1948):81-117. *T*

A great deal of information is presented here about the frequency, duration, and rhythms of infant sucking patterns as measured with an apparatus inserted into the bottle and attached to a polygraph. The entire sucking pattern, from beginning to end per feeding, of some one hundred children was analyzed for age, sex, and individual differences as well as for relationships to physical illnesses. Sample findings are that lower frequencies are seen in the youngest babies; that the fastest sucking occurs in "irritable" babies and infants with intestinal disorders; and that there are marked individual differences in pattern.

41   **Barker, Larry L., and Collins, Nancy B.** "Nonverbal and Kinesic Research." In *Methods of Research in Communication*, edited by P. Emmert and W. D. Brooks, pp. 343-372. Boston: Houghton Mifflin Company, 1970. *D N*

Focuses on Birdwhistell's approach to kinesic research and the VID-R computer system developed by Ekman et al. for movement research.

42   **Barker, Roger G.,** ed. *The Stream of Behavior: Explorations of Its Structure and Content.* New York: Appleton-Century-Crofts, 1963. 352 pp. *T*

A series of chapters dealing with problems of defining behavioral units and analyzing the stream of behavior and social interaction in vivo. Particularly relevant to body movement research for its explorations in behavior pattern analysis and for the ways that nonverbal communication is described within the behavior stream.

43   **Barlow, Wilfred.** "Anxiety and Muscle Tension." In *Modern Trends in Psycho-*

somatic Medicine, edited by D. O'Neill, pp. 285-309. New York: Paul B. Hoeber, 1955. *G P*

Techniques of muscular reeducation similar to Alexander's are described and evaluated. "Postural homeostasis" is discussed, and photographs before and after reeducation are presented along with notes on the behavior of the music student subjects.

44 **Barratt, Ernest S.** "Perceptual-Motor Performance Related to Impulsiveness and Anxiety." *Perceptual and Motor Skills* 25 (1967):485-492. *G*

Subjects were divided into four groups on the basis of high or low scores on tests of anxiety and impulsivity and then given four perceptual-motor tasks to perform. High impulsivity-low anxiety subjects were consistently less efficient, thus suggesting that impulsiveness is related "more to motor control than to sensory discrimination or cognition."

45 **Bartenieff, Irmgard, and Davis, Martha.** "Effort-Shape Analysis of Movement: The Unity of Expression and Function." Unpublished monograph. Bronx, N.Y.: Albert Einstein College of Medicine, 1965. 71 pp. *D N T* (Copies from Dance Notation Bureau, 8 East 12th St., New York 10003)

A comprehensive initial exposition of Rudolf Laban's effort-shape analysis and notation of movement, with illustrations, a history of its development, and discussion of how body movement patterns can be interpreted psychologically with Laban's theory of movement behavior. In a chapter on neurophysiological mechanisms and visible aspects of body movement, effort-shape analysis is posited to describe the "graded motor response." Extensive discussion and charts tracing early development of individual movement characteristics, posture-locomotion, prehensile activity, and visual-motor coordination are presented with an effort-shape analysis of their functional significance. From pilot psychiatric research, "movement case studies" of hospitalized patients and notes on movement behavior observed in a family therapy session are described.

46 **Bartenieff, Irmgard, and Davis, Martha.** "An Analysis of the Movement Behavior within a Group Psychotherapy Session." Presented at the Conference of the American Group Psychotherapy Association, Chicago, 1968. 28 pp. *D G* (Copies from Dance Notation Bureau, 8 East 12th St., New York 10003)

A videotape of a group psychotherapy session was studied using Laban's "effort-shape" analysis of movement. Notes were made of each member's body attitude, postural shifts, body parts moved, "effort dynamies," and spatial characteristics of movement. An analysis of movement characteristics observed over six periods in the session demarcated by changes in subgrouping was presented. Interpretation of body movement characteristics shared by the group as to group mood and kinds of relatedness were presented, with focus on the nonverbal aspects of a specific encounter. Also included are descriptions of the movement "style" of each member and how this is interpreted.

47 **Bartenieff, Irmgard; Davis, Martha; and Paulay, Forrestine.** *Four Adaptations of Effort Theory in Research and Teaching.* New York: Dance Notation Bureau, forthcoming.

Contains a biographical note about Rudolf Laban, a chapter on the logic and consistency of his effort-shape analysis of body movement, and a chapter on dance aesthetics drawn from Laban's early writing; discussion of the potential of

Laban's theories of space harmony for movement retraining and the relation of space harmony to neurophysiological concepts; and a chapter on the choreometrics project of Bartenieff, Paulay, and Lomax, a cross-cultural study of dance.

48   Bartenieff, Irmgard, and Paulay, Forrestine. "Choreometrics Profiles." In *Folk Song Style and Culture* by Alan Lomax, pp. 248-261. Washington, D.C.: American Association for the Advancement of Science, Publication no. 88, 1968. *G*

Movement profiles or graphs based on choreometrics analysis of anthropological films are presented, together with verbal descriptions of the dance and work styles of nine different cultures. The cultures described according to their movement styles are: Eskimo, Iroquois, Australian Aboriginal, New Guinea Maring, Samoan, Ellice Islander, African, Indian from Malabar, and English Morris.

49   Bartenieff, Irmgard, and Paulay, Forrestine. "Dance as Cultural Expression." In *Dance: An Art in Academe*, edited by M. Haberman and T. G. Meisel, pp. 23-31. New York: Teachers College Press, Columbia University, 1970.

A clear, concise description of choreometrics research on cultural dance and work styles and the implications of the projects' hypotheses and findings for American dance education. Specific movement styles—those of the Eskimo, Iroquois, West African, and Ellice Islander—are summarized as various modes of adjustment to their environment, and the conflict between African and Western European movement styles is described in the body movement of an American Negro dancer.

50   Bastock, Margaret. *Courtship: An Ethological Study.* Chicago: Aldine Publishing Company, 1967. 220 pp. *D T*

The courtship and mating behavior—interaction, displays, movements—of various fish, birds, and arthropods are described and illustrated. The evolution and social function, as well as hormonal and nervous mechanisms, of courtship are discussed.

51   Bateson, Gregory; Birdwhistell, Ray L.; Brosin, Henry W.; Hockett, Charles; and McQuown, Norman A., eds. *Natural History of an Interview*, nos. 95-98 (ser. 15), Microfilm Collection of Manuscripts in Cultural Anthropology. Chicago: University of Chicago Library, 1971.

Although this work has never been published, it is frequently cited in reference lists, and its importance has been recognized for several years. A collection of chapters by different authors, it represents a historic collaboration among linguists, anthropologists, and psychiatrists such as Frieda Fromm-Reichmann on the nature of group interaction and communication. Two chapters by Ray L. Birdwhistell have been published in his *Kinesics and Context*. The full work will not be published, but a microfilm or xerox copy of the manuscript is available from Joseph Regenstein Library, Photoduplication Department, University of Chicago, 1100 East 57th Street, Chicago, Ill. 60637.

52   Bateson, Gregory, and Mead, Margaret. *Balinese Character: A Photographic Analysis.* New York: New York Academy of Science, Special Publications, vol. 2, 1942. 277 pp. *P*

When first published, this book was an "experimental innovation" using photographs to record and describe the Balinese—"the way in which they, as living persons, moving, standing, eating, sleeping, dancing and going into trance embody that abstraction" called culture (p. xii). It includes an extraordinary collection of 100 groups of photographs of the Balinese in a multitude of settings, transactions, and body expressions, accompanied by behavioral descriptions by the anthropologists and their analysis of the behavior and characteristics of the Balinese.

53   Baxter, James C.; Winters, Elaine P.; and Hammer, Robert E. "Gestural Behavior During a Brief Interview as a Function of Cognitive Variables." *Journal of Personality and Social Psychology* 8 (1968):303-307.

Forty young men and women were asked to talk for five minutes on a familiar topic of their choice. An observer counted the number of movements made, the number of vigorous movements, and the number of gestures toward and away from the body, while a stabilometer fixed to the chair also recorded the number of movements. Subjects had completed a form testing "individual differences in verbal category differentiation." The researchers found that the more verbally well-differentiated person was more active, more vigorous, and more extended in his movements.

54   Baxter, Mildred F. "An Experimental Study of the Differentiation of Temperaments on a Basis of Rate and Strength." *American Journal of Psychology* 38 (1927):59-96. G T

Subjects were given motor and mental tests of reaction speed and strength and questionnaires about aspects of their temperament. The speed and strength tests did not clearly define four types of temperament, and correlation with temperament was not substantiated.

55   Bayley, Nancy. "The Development of Motor Abilities during the First Three Years." *Monographs of the Society for Research in Child Development* 1 (1935): 1-26. G T

Presents the results of the California Infant Scale of Motor Development, derived from repeated testing of sixty-one children from ages one month to three years, along with a description of the testing procedures, materials, scoring, and a sample of the California Infant Scale itself. The scale, consisting of a series of motor tasks for the child (one task per age interval), is not so reliable as the mental scales. Motor scores show a correlation with mental scores in the first fifteen months, but afterward this decreases. Early motor scores are not predictive of later ones. No correlation was found between motor scores and parents' level of education. Body build appears to have little relation to motor ability in infancy. The monograph includes a number of observations concerning the rate and pattern of motor development.

56   Bayley, Nancy. "Comparisons of Mental and Motor Test Scores for Ages 1-15 Months by Sex, Birth Order, Race, Geographical Location and Education of Parents." *Child Development* 36 (1965):379-411. G T

Bayley's Scales of Mental and Motor Development were given to 1,409 infants, ages one to fifteen months, from twelve cities, a population representative of the United States. No score differences were found between boys and girls, first-born infants as compared with later-born, education level of either parent, or geographical location. The only difference found was that Negro babies performed better on the motor scales than did whites; possible interpretations of this are discussed.

57   Bayley, Nancy, and Jones, Harold E. "Environmental Correlates of Mental and Motor Development: A Cumulative Study from Infancy to Six Years." *Child Development* 8 (1937):329-341. G T

Sixty-one children of varying socioeconomic levels were given mental and motor tests regularly from one month to six years of age. Results of the motor test (California Infant Scale of Motor Development) showed very low correlations between motor development and environmental factors, some positive correlations with mother's education level between fifteen and thirty-three months, and no correlation between motor development and father's education or family income level.

58   **Beaumont, Cyril W.** *A Bibliography of Dancing.* London: Dancing Times, 1929. 228 pp. [New York: Benjamin Blom]

An annotated bibliography of books on dance in the British Museum Library up to 1929, many of them rare and probably now out of print; extensively indexed by subject.

59   **Beaumont, Cyril, ed., and Leslie, Serge, annotator.** *A Bibliography of the Dance Collection of Doris Niles and Serge Leslie.* 2 vols. London: C. W. Beaumont. 1966, 1968. 594 pp.

A beautifully annotated, well-indexed two-volume bibliography of some 2,000 titles, most of which are dance biographies and synopses; but a number deal with aesthetics, notation, and ethnic and historical dance.

60   **Beckson, Gershon, and Davenport, Richard K.** "Stereotyped Movements of Mental Defectives. I, Initial Survey." *American Journal of Mental Deficiency* 66 (1962):849-852. *G T*

A study of the prevalence and interrelationship of repetitive movements such as rocking, scratching, and holding the hands before the eyes of seventy-one institutionalized mentally retarded men. Stereotyped behaviors were found to be related to I.Q. and length of hospitalization; also, blind patients showed more stereotyped movement than the sighted men did.

61   **Beier, Ernst G., and Stumpf, John.** "Cues Influencing Judgment of Personality Characteristics." *Journal of Consulting Psychology* 23 (1959):219-225. *T*

A study of how constant judges' perception of certain positive and negative personality traits are when they are presented cumulatively with voice, then gestures, and then facial expression and finally with the subject interacting with them. Results were mixed.

62   **Beigel, Hugo G.** "The Influence of Body Position on Mental Processes." *Journal of Clinical Psychology* 8 (1952):193-199.

Male subjects were presented TAT cards and provocative letters while under hypnosis, and each card or letter was responded to as the subject stood, sat, and reclined. Analysis of the responses indicates that when standing the subjects' associations were more active, goal-directed, and decisive and emotions were expressed more vigorously; whereas when reclining they were more complacent, passive, and reflective. The reactions while sitting were a mixture of the two types of characteristics, but were closer to the standing reactions.

63   **Belknap, S. Yancy,** compiler. *Guide to Dance Periodicals, Vol. I: 1931-1935.* Gainesville: University of Florida Press, 1959. 123 pp.

Titles are listed by author and subject, including references to aesthetics and ethnological dance. Nine additional volumes cover the years up to 1963. Later volumes are published by The Scarecrow Press, Metuchen, New Jersey.

64   **Bell, Sir Charles.** *The Anatomy and Philosophy of Expression as Connected with the Fine Arts.* 4th ed. London: John Murray (Publishers), 1847. 275 pp. *D* [Farnborough, Eng.: Gregg International Publishers, 1971]

Written in 1806, this is an exploration of the aesthetics, psychology, and anatomy of bodily expression and structure in humans and animals. There is an analysis of racial and cultural differences in head structure; discussion of facial muscles and the expressions to which they are related; and particular note of the expressiveness of eye movements. Facial and body expression of emotions is analyzed, with examples from art and literature and from the author's own observations. Numerous expressions are illustrated, described, and assessed psychologically and anatomically in this historically important work.

65 **Bender, Lauretta, and Boas, Franziska.** "Creative Dance in Therapy." *American Journal of Orthopsychiatry* 11 (1941):235-244. *D*

Therapy sessions using dance and percussion instruments with small groups of children in a psychiatric hospital are vividly described. What held the group together, how the children dramatized their conflicts, and how they interrelated are discussed. A theoretical rationale for using dance therapeutically is presented.

66 **Bender, Lauretta, and Helme, William H.** "A Quantitative Test of Theory and Diagnostic Indicators of Childhood Schizophrenia." *AMA Archives of Neurology and Psychiatry* 70 (1953):413-427. *T*

An outline of the nature and symptoms of childhood schizophrenia is presented through data from nurses' notes, neurological examination, drawings, psychiatric interviews, etc. (much of it on motor and body-image disturbances), collected for groups of schizophrenic and nonpsychotic children. Among other findings, the childhood schizophrenic group showed a dominance of "tonic-neck-reflex motility" and problems of "boundary maintenance in the psychological world."

67 **Benesh, Rudolph and Joan.** *An Introduction to Benesh Dance Notation.* London: A. & C. Black, 1956. 48 pp. *D N*

One of the most widely developed and used dance notations, this system has a horizontal staff with divisions for parts of the body, symbols for level and for front and behind of the body, relative placement on the staff for positions in space, and lines indicating the path of the movement. This book describes the basic principles of the system as applied to ballet.

68 **Bentley, Eric.** "The Purism of Etienne Decroux." In *In Search of Theater,* pp. 174-184. New York: Vintage Books (Random House), 1959.

Using interviews with Decroux, the author summarizes this mime artist's aesthetics. Fascinating contrasts are made between mime and dance and between mime and drama. Also of note are descriptions of how the mime achieves certain effects.

69 **Berdach, Elsie, and Bakan, Paul.** "Body Position and the Free Recall of Early Memories." *Psychotherapy: Theory, Research and Practise* 4 (1967):101-102. *T*

An experimental study of twenty-one subjects, which shows that earlier memories are recalled while lying down.

70 **Berger, Milton M.** "Nonverbal Communications in Group Psychotherapy." *International Journal of Group Psychotherapy* 8 (1958):161-178.

Includes a group psychotherapist's clinical observation of various actions, postures, and manners of moving among patients and, based on the patients' comments and histories, what their emotional and psychodynamic significance appeared to be.

71 **Bergler, Edmund.** *Laughter and the Sense of Humor.* New York: Intercontinental Medical Book Corporation, 1956. 297 pp. *T* [New York: Grune & Stratton, 1967]

Includes a chapter on the many theories about the nature of laughter and another on the development of smiling and laughter in children.

72   **Berk, Robert L.** "A Comparison of Performance of Subnormal, Normal, and Gifted Children on the Oseretsky Tests of Motor Proficiency." Ph.D. dissertation, Boston University, 1957. 280 pp. (Datrix order no. 22115)

The author's revision of the Lincoln Adaptation to the Oseretsky Scale of Motor Proficiency was administered to gifted, normal, and below-normal children matched by age, sex, and race. None had motor disturbance due to brain damage. A low positive correlation between the motor tests and the Stanford-Binet I.Q. test was found, primarily owing to differences between the subnormals and the normals. There was little difference between the motor performance of the normals and the gifted children except in the Simultaneous Voluntary Movement Test.

73   **Bernal y del Río, Victor.** "On Psychotherapy." *Boletín de la Asociación Médica de Puerto Rico* 60 (1968):553-559.

A discussion of psychotherapy practice in Puerto Rico, based on supervision of residents and focusing on nonverbal interaction. The author discusses what he calls "microechopraxia" (similar movement in patient and therapist) and its importance in transference and countertransference.

74   **Bernal y del Río, Victor.** "Imitation, Identification and Identity (Remarks on Transcultural Observations)." Presented at the Herman Goldman International Lecture Series, New York Medical College, May 1971. 9 pp. (Copies from author: 300 Franklin D. Roosevelt Avenue, San Juan, Puerto Rico)

In discussing therapy practice with upper- and lower-income Puerto Rican patients and "transient" and "replanted" continentals, the author describes differences in handshaking behavior and contrasts between the patients' body movements in therapy sessions versus outside the therapy situation in Puerto Rico. References to the hyperkinetic seizures of the "Puerto Rican syndrome" are cited.

75   **Bhadbury, Manjulika.** *The Art of the Hindu Dance.* Calcutta: S. K. Chatterjee, 1945. 275 pp. *D P*

The origins, history, and performance of Indian Hindu dance are discussed, with extensive description, illustration, and interpretation of the meaning of the hand gestures, postures, and footwork involved.

76   **Bieber, Irving.** "Grasping and Sucking." *Journal of Nervous and Mental Disease* 91 (1940):31-36.

An analysis of the relationship between grasping and sucking as observed in infants and certain adult patients with cerebral disease.

77   **Bills, Arthur G.** "The Influence of Muscular Tension on the Efficiency of Mental Work." *American Journal of Psychology* 38 (1927):227-251. *G T*

Investigating the effects of muscle tension (produced by squeezing dynamometers) on various types of mental work, such as learning a list of nonsense syllables, the author found that tension did increase efficiency and that, if speed of learning is the criterion, tension increased efficiency with practice and when subjects were growing fatigued. Possible neural or physiological explanations of the experimental results are discussed.

78   **Birdwhistell, Ray L.** *Introduction to Kinesics.* Louisville, Ky.: University of Louisville Press, 1952. *N*

As Birdwhistell's first presentation of a notation for body motion and the nature of his kinesics research, this is a historically important publication which foreshadowed a "renaissance" in body movement research. Currently it is available in its entirety only on microfilm from University Microfilms Inc., 313 North First St., Ann Arbor, Michigan; but part of it is reprinted in Birdwhistell's *Kinesics and Context* (see below).

79 **Birdwhistell, Ray L.** "The Kinesic Level in the Investigation of the Emotions." In *Expressions of the Emotions in Man,* edited by P. H. Knapp, pp. 123-139. New York: International Universities Press, 1963. (Copies from author, Eastern Pennsylvania Psychiatric Institute, Philadelphia, Pa.)

Birdwhistell argues that the study of affect as well as of individual differences depends on an understanding of learned cultural patterns of communication. He questions dichotomies such as ideation-emotion and notions that words relate to the cognitive and movement relates to the affective because it is "closer to nature." He discusses the relation of kinesics to physiology and presents concepts such as "motion qualifiers" and "expectancy qualifiers" as potentially useful in affect studies. Head-nodding patterns are described to illustrate motion qualifiers.

80 **Birdwhistell, Ray L.** "Communication Without Words." In *L'Aventure Humaine.* Encyclopédie des Sciences de L'Homme, vol. 5 pp. 157-166. Geneva: Editions Kister, 1968. *D N* (Copies from author, see above)

Birdwhistell makes many of the theoretical points of this paper elsewhere, and the second half is reprinted in *Kinesics and Context.* However, the critique here of studies of emotions is particularly clear, and there is an interesting section on the context and culture-boundness of gestures that has not been reprinted. Birdwhistell describes the evolution of his kinesics research as an approach similar to that of the structural linguist.

81 **Birdwhistell, Ray L.** "Kinesics, Inter- and Intra-Channel Communication Research." In *Studies in Semiotics,* edited by T. A. Sebeok, International Social Science Council, vol. 7, pp. 9-26. Paris: Mouton Publishers, 1968. (Copies from author, see above)

Reiterating his approach to communication research, Birdwhistell argues that the social level is the proper level at which to study it, not a psychological or physiological one, and that limiting the research to abstracted bits of behavior or single "channels" such as speech activity may facilitate research but is in itself distorting and incomplete. If communication, and body movement as one channel within it, is culturally coded, ultimately the code must be deciphered at the multi-channeled, social level. He illustrates his arguments with a discussion of how smiling behavior is culturally learned; this section is reprinted in the author's *Kinesics and Context* (see below).

82 **Birdwhistell, Ray L.** "Nonverbal Communication in the Courtroom: What Message Is the Jury Getting?" In *Persuasion: The Key to Damages,* pp. 189-204. Ann Arbor, Mich.: The Institute of Continuing Legal Education, 1969. (Copies from author, see above)

An informal discussion of dualistic notions such as that the upper body and speech equals what is clean and good, whereas the lower body is associated with affect and what is to be controlled. To illustrate how words are a very small part of communication and that this has many ramifications for law, Birdwhistell illustrates how we inform others who we are and where we are from by variations in body appearance and movement, in this case focusing on facial characteristics and head carriage.

83 **Birdwhistell, Ray L.** "Part I, Learning to Be a Human Body." In *Kinesics and Context,* pp. 3-62. Philadelphia: University of Pennsylvania Press, 1970. *D*

Through his pioneering research in "kinesics," the term he coined for cultural patterns of body motion communication, Birdwhistell is at the forefront of the current interest in and investigation of body movement. The work of several other researchers, notably Condon, Scheflen, Kendon, and Loeb, is directly related to his; and many other researchers are indirectly influenced by him. *Kinesics and Context* is a compilation of his articles and speeches on kinesics from the early 1950's until 1969. In Part I he argues that the child assimilates a complex hierarchy of cultural communication patterns which make him a predictable and, therefore, accepted member. The nature of this assimilation is discussed and is illustrated with descriptions of family interaction, cultural variations in the smiling response, sex differences in posture, and facial expressions and "pathological" communication patterns observed in two films of mother-child interaction. Birdwhistell eschews the search for universal expressions, instinctual behavior, and individual idiosyncrasies in body movement and argues that kinesic patterns are cultural and context-specific and require natural history research techniques and methods similar to those used by linguists.

84   **Birdwhistell, Ray L.** "Part II, Isolating Behavior." In *Kinesics and Context,* pp. 65-91. Philadelphia: University of Pennsylvania Press, 1970. *D*

In this theoretical section of the book, Birdwhistell argues that communication is not "unilateral transmission" or simple back-and-forth verbal dialogue but a continuous process involving combinations of linguistic, kinesic, tactile, olfactory, and other "channels." Motions may be made up of discrete units that, like sounds, combine into "sentences" and so forth, but their meaning varies according to context, how they are done, who does them, etc. The communication pattern and its rules should be studied on the social level before individual variations can be understood. Birdwhistell discusses what aspects of kinesics research occur "coextant" or congruent with speech structures, the nature of redundancy in communication, and the "integrational" (maintenance, regulation, clarification, and so on) aspects of communication.

85   **Birdwhistell, Ray L.** "Part III, Approaching Behavior." In *Kinesics and Context,* pp. 95-143. Philadelphia: University of Pennsylvania Press, 1970. *D N*

Birdwhistell hypothesizes that body motion is composed of a limited number of elements, combinations and rules for usage, and he lists the fifty to sixty head, face, and eye kinesic units ("kinemes") he has found in American movement. In a discussion of how his research developed, he talks about formal sign language. Kinemes combine according to his framework into larger units called "kinemorphs," which in turn combine into "kinemorphic constructions." He illustrates certain relationships between speech and movement, particularly some fascinating formulations about kinesic "markers" as these correspond to "parts of speech" (e.g., "pluralization" markers, in which a sweep of the body part appears to occur with plural nouns or pronouns; or "tense" markers, in which a movement backward occurs with past tense verbs and a movement forward occurs with future tense verbs). In addition, there are illustrated examples of the relation between linguistic stress and kinesic stress and juncture. In this section Birdwhistell also discusses "parakinesics" and its possible functions in interaction and relation to "formal" kinesic constructions.

86   **Birdwhistell, Ray L.** "Part IV, Collecting Data: Observing, Filming, and Interviewing." In *Kinesics and Context,* pp. 147-170. Philadelphia: University of Pennsylvania Press, 1970. *N*

The advantages and limitations of film, videotape, and photographs for research and how selections made by the cameraman and the judgments made by untrained observers (to establish observer reliability) reflect their culturally learned biases, and not necessarily what is true about the behavior observed. Given the redundancy and repetition in behavior, Birdwhistell states that a small segment of a film may be sufficient if its context is documented. He then discusses notions of tempo and sequence, giving an example of an adolescent courtship pattern where to be "fast" is to leave out a step, not to perform the stages in a short time. Two chapters present detailed and very technical analysis of American head nodding and another deals with eyebrow movement, together with their relation to speech and sustaining interaction. The section concludes with an early paper on training interviewers in the field to be good observers of body movement.

87   **Birdwhistell, Ray L.** "Part V, Research on an Interview." In *Kinesics and Context,* pp. 173-251. Philadelphia: University of Pennsylvania Press, 1970. *D N T*

This section, composed of Birdwhistell's two chapters in *The Natural History of an Interview* (see 51) is particularly valuable because it seems to be his most comprehensive and best-written exposition of kinesics and because the often-quoted original source has not yet been published. Birdwhistell begins by introducing the concept of kinesics with an elaborate analysis and notation of the behavior of two soldiers thumbing a ride and a driver who passes them up. He goes on to review the literature on body movement and the major influences on his own development. He succinctly states the premises underlying his approach, his critique of other approaches, and his method for abstracting kinesic units and determining their communicational significance. A number of kinesic units and terms are defined: kine, kinemorph, kinemorphic construction, stance, motion qualifiers, action modifiers, interaction modifiers, and motion markers. Some terms, such as body-base and body-set, are illustrated with examples such as the description of the behavior of a young executive and his boss or the different ways that illness is expressed in two Kentucky subcultures. Finally, in a very technical and detailed linguistic-kinesic analysis of eighteen seconds of a film, Birdwhistell illustrates his methods, concepts, and notation.

88   **Birdwhistell, Ray L.** "Appendixes." In *Kinesics and Context,* pp. 255-304. Philadelphia: University of Pennsylvania Press, 1970. *N*

These appendixes include a section on Birdwhistell's system of notation reprinted from his hard-to-obtain *Introduction to Kinesics* (1952), as well as sample recordings and what appear to be later additions and modifications of kinesic notation. In this system the body is divided into eight parts, with each "kine" of face, head, trunk, etc., represented by a discrete notation. The notation is composed of letters, numbers, arrows, symbols, and simple drawings—some of these having an apparent relation to the body movement or position represented, and some arbitrarily assigned.

89   **Birns, Beverly.** "Individual Differences in Human Neonates' Responses to Stimulation." *Child Development* 36 (1965):249-256. *T*

Four different stimuli were applied to neonates between two and five days old, and trained observers rated the degree of movement, from "no response" to "intense overall activity." Throughout four sessions the babies showed significant individual consistency and reacted at a characteristic level no matter what the stimulus.

90   **Bishop, Alison.** "Use of the Hand in Lower Primates." In *Evolutionary and Genetic Biology of Primates,* vol. 2, edited by J. Buettner-Janusch, pp. 133-325. New York: Academic Press, 1964. *D G P T*

Includes detailed description and analysis of the hand's evolution and function in locomotion, feeding, and social behavior.

91   **Blackman, Bernice.** "Three Studies in Hyperactivity: III, A Comparison of Hyperactive and Non-Hyperactive Problem Children." *Smith College Studies in Social Work* 4 (1933):54-65. *T*

The physical characteristics, intelligence, health histories, and developmental patterns, including onset of walking, of fifty hyperactive children are compared with a control group.

92   **Blank, Marion.** "A Focal Periods Hypothesis in Sensorimotor Development." *Child Development* 35 (1964):817-829. *T*

Forty infants were tested with the Griffiths Mental Development Scale, and their manipulation of toys was systematically observed and rated by the author at about twenty weeks, forty weeks, and, for some, seventy-two weeks of age. The evaluation of hand skills was the best indicator of overall developmental level at twenty weeks but not later. This and some unsubstantiated observations suggest that there are diverse focal periods of development for prehension skills, locomotor performance, gross motor activity, and vocalization.

93   **Blanton, Margaret Gray.** "The Behavior of the Human Infant during the First Thirty Days of Life." *Psychological Review* 24 (1917):456-483.

Based on observation of many neonates, the author describes onset, incidence, and types of many reactions, including eye movements, facial expressions, reflexes, bodily reactions to varying stimuli, and head, limb, and total body movements.

94   **Blasis, C.** *The Code of Terpsichore: The Art of Dancing.* Translated by R. Barton. London: Edward Bull, 1830. 548 pp. *D*

Subtitled "Its Theory and Practice, and A History of Its Rise and Progress from the Earliest Times." A very interesting nineteenth-century view of Greek and Roman dancing, ethnic and "private" (social) dances of the time, ballet, pantomime, and symbolic or natural gestures of different countries. See also Carlo Blasis, *Elementary Treatise upon the Theory and Practice of the Art of Dancing,* translated by M. S. Evans, 3d ed., New York: Dover Publications, 1968.

95   **Blatz, William E., and Millichamp, Dorothy A.** "The Development of Emotion in the Infant." *University of Toronto Studies, Child Development Series* 4 (1935): 1-44. *G T*

A study of five infants from age one to twenty-four months in their home setting. "Emotional episodes"—here referring primarily to negative or distressed reactions —are defined in terms of observable behavior and physical movement. Such behavior becomes less frequent, but more complex and differentiated, with age. The situations evoking the reactions are analyzed, primarily as forms of thwarting approach and withdrawal attitudes.

96   **Blatz, William E., and Ringland, Mabel Crews.** "A Study of Tics in Pre-School Children." *University of Toronto Studies, Child Development Series* 3 (1935):1-58. *T*

A quantitative study of tics in fifty-seven nursery school and kindergarten children. "Tic" is very broadly defined here as everything from frowning or rubbing the eyes to swaying of the body. The monograph has an excellent review of early medical and psychological literature on tics and "nervous habits." Frequency of specific actions according to age and situation is given, the data indicating that there are individual characteristics but no sex differences, and that there is an increase of such nervous reactions with age and in situations of greater physical restraint.

97 **Blauvelt, Helen.** "Further Studies on Maternal-Neonate Interrelationships." In *Group Processes: Transactions of the Third Conference*, edited by B. Schaffner, pp. 195-217. New York: Josiah Macy, Jr. Foundation, 1957.

The movements of mother, baby and nurse are sensitively described from films of neonates during feeding, and the behavior observed is analyzed by members of the conference.

98 **Blazer, John A.** "Leg Position and Psychological Characteristics in Women." *Psychology* 3 (1966):5-12. *T*

The characteristic leg positions of 1,000 Southern white women were compared with their personality characteristics as assessed through the EPPS, Study of Values, and WAIS.

99 **Blest, A.D.** "The Concept of Ritualisation." In *Current Problems in Animal Behaviour,* edited by W. H. Thorpe and O. L. Zangwill, pp. 102-124. Cambridge: Cambridge University Press, 1961.

A discussion and review of prior concepts of ritualization and signal movements in lower animals, their evolution, definition, and function.

100 **Bleuler, Eugen.** *Dementia Praecox or the Group of Schizophrenias.* Translated by J. Zinkin. New York: International Universities Press, 1950. 548 pp. [Latest edition: 1966]

Under "catatonic symptoms" (pp. 180-206), Bleuler describes numerous motor disturbances of the schizophrenic patients of his time: catalepsy, echopraxia, akinesia, waxy flexibility, catatonic hyperkinesis, stereotypies, mannerisms, mimetic expressions, and impulsiveness. In another section (pp. 441-461), he proposes theories concerning their origins and psychological interpretation. There are also descriptions of body movements and facial expressions occurring in manic and depressive states.

101 **Block, Helen.** "The Influence of Muscular Exertion upon Mental Performance." *Archives of Psychology* 202 (1936):1-49. *G P T*

An example of studies of the effects of sustained physical pressure during mental task, this one shows wide. inconsistent individual differences in fifteen male subjects and no clear relationship between exertion and mental performance.

102 **Blount, W. P.** "Studies of the Movements of the Eyelids of Animals: Blinking." *Quarterly Journal of Experimental Physiology* 18 (1927):111-125. *G*

Data on the blinking patterns of a wide range of zoo and domestic animals, from birds and amphibians to lions and elephants.

103 **Boas, Franziska,** ed. *The Function of Dance in Human Society.* N.p.: F. Boas, 1944. 52 pp. *P* (To be reprinted by Dance Horizons Press, Brooklyn, N.Y.)

Based on a seminar on the "relationship between dance and the way of life," there are chapters by F. Boas on the Kwakiutl American Indians, by G. Gorer on dance forms of African communities, by H. Courlander on dance-drama in Haiti, and by C. Holt and G. Bateson on the function of dance in Bali.

104 **Bolwig, Niels.** "A Study of the Behaviour of the Chacma Baboon." *Behaviour* 14 (1959):136-163. *D*

In this very interesting description of the social behavior of captive and wild baboons, made from field notes and film analysis, there are detailed illustrated analyses of the baboon's movement patterns and/or facial expressions in situations of threat, attack and submission, play, grooming, and mating. Bolwig also

describes the expression of conflicting emotion and patterns of infant-mother interaction.

105 **Bolwig, Niels.** "Facial Expression in Primates, with Remarks on a Parallel
Development in Certain Carnivores (A Preliminary Report on Work in Progress)".
Behaviour 22 (1963-64):167-192. *D P*

A number of monkeys and apes were observed and photographed in various
contexts. The evolution of facial muscles from primitive primates to man is described, based on dissection, and is illustrated with anatomical drawings and compared with the facial anatomy of dogs. The author then describes primate behaviors interpreted as indicating various emotions (joy, happiness, fear, sadness,
anger, affection, etc.), the contexts in which they occur, which animals do them,
and, especially, the pattern of facial expressions associated with the emotion.
Photographs and drawings illustrate how humans and different primates perform
the facial expressions—showing that the higher the primate, the more developed
the facial expression, with origins detectable in specific actions such as biting,
screaming, and sucking. The author notes that primates tended by humans may
show more exaggerated expressions than do animals in the wilds.

106 **Boomer, Donald S.** "Speech Disturbance and Body Movement in Interviews."
Journal of Nervous and Mental Disease 136 (1963):263-266.

The number of "nonpurposive movements" and different types of speech disturbance. were tabulated in thirty-nine segments from one patient's psychotherapy
sessions. A relationship was found between the body movement and the "filled
pause" (e.g., "ah" or word repetition) type of speech disturbance.

107 **Boomer, Donald S., and Dittmann, Allen T.** "Speech Rate, Filled Pause,
and Body Movement in Interviews." *Journal of Nervous and Mental Disease*
139 (1964):324-327. *T*

The authors hypothesize that speech rate, "filled pauses," and number of
body movements increase under "emotional disturbance"—meant here as stress
—with an increase in filled pauses and body movement seen as secondary effects
of changes in speech. Films were analyzed of eight young male subjects talking
on a subject of their choosing, either with no instructions, with the command not
to hesitate or pause, or with instructions to speak without using any words with
an "L" in them. With the no-pauses situation, the subjects' movement and filled
pauses decreased; in the stressful, no-L's condition, subjects did have more filled
pauses but not more movement, thus refuting the hypotheses.

108 **Boring, E.G., and Titchener, E.B.** "A Model for the Demonstration of Facial
Expression." *American Journal of Psychology* 34 (1923):471-485. *D*

This curious paper describes a teaching aid which the authors use to demonstrate their psychology lectures on expression of emotion. The aid is a profile
based on Piderit's "geometry of expression," which has pieces for various eye,
nose, cheek, and mouth aspects that can be fitted together to make different facial
expressions. As the authors present a long list of the possible variations and how
these are interpreted, they, in effect, summarize interpretations of facial expression
made by Darwin and Wundt.

109 **Boucher, Jerry D.** "Facial Displays of Fear, Sadness and Pain." *Perceptual
and Motor Skills* 28 (1969):239-242.

In two different experiments, subjects showed high agreement in judging specific
effects from facial expression, in this case discriminating between fear, sadness,
and pain.

110   **Boucher, Jerry D. and Ekman, Paul.** "A Replication of Schlosberg's Evaluation of Woodworth's Scale of Emotion." Presented at the Western Psychological Association Convention, n.p., June 1965. 9 pp. *D* (Copies from P. Ekman, Studies in Nonverbal Behavior, 1405 Fourth Avenue, San Francisco, Calif. 94122)

Part of a study of what emotions are conveyed in facial expression. Sixty photographs of people's faces taken during stress interviews were presented to students to be rated by one group according to Woodworth's six categories of emotion and by another according to Schlosberg's pleasant-unpleasant and rejection-attention dimensions. The authors found a lower correlation between the two ratings than had been reported. They discuss the efficacy of the emotion scales and the complicated nature of making judgments of facial expression.

111   **Bowlby, John.** *Attachment and Loss,* Vol. 1, *Attachment.* New York: Basic Books, 1969. 428 pp.

The entire book is devoted to the analysis of "attachment behavior" and the mother-child relationship; however, Chapter 11, entitled "The Child's Tie to His Mother," is particularly relevant here for the behavioral analyses of mother-child interaction in humans and primates with reference to touch, proximity, and five "primary" patterns of behavior contributing to attachment: "sucking, clinging, following, crying, and smiling."

112   **Boyd, John.** "Comparison of Motor Behavior in Deaf and Hearing Boys." *American Annals of the Deaf* 112 (1967):598-605. *T*

Based on motor tests from the Oseretsky Test of Motor Proficiency and the Van Der Lugt Psychomotor Test Series for Children, the deaf schoolchildren did significantly poorer than the hearing children in static equilibrium tests and locomotor coordination (after eight years of age). No differences were found between the deaf boys with differing etiologies, except in laterality and speed tests.

113   **Boynton, M. Adelia, and Goodenough, Florence L.** "The Posture of Nursery School Children During Sleep." *American Journal of Psychology* 42 (1930): 270-278. *T*

The postures of fifty-six nursery-age children were recorded during their afternoon naps. There were no marked differences in age or sex. The children changed posture on an average of once in twenty-five minutes; spent most time on their right sides, least on their backs; and moved least during the early stages; the more uniform their sleep postures, the more quickly they fell asleep.

114   **Braatöy, Trygve.** "Psychology vs. Anatomy in the Treatment of 'Arm Neuroses' with Physiotheraphy." *Journal of Nervous and Mental Disease* 115 (1952): 215-245. *D G P*

In a disarmingly simple and clear way, Braatöy proceeds from routine physical examination of arm tension and an anatomical analysis of the muscles involved to the relation of certain postures and arm tensions to the startle response and defensive reactions. He shows how arm tension resulting from occupational strains or from momentary shyness can be distinguished from chronic protective attitudes with psychogenic origins. With experimental data and clinical examples, he postulates relationships between localized muscle tensions, fixed postures and their psychological origins, and defensive functions. Noting that physiotherapists and medical doctors often overlook the psychological significance of the tensions they treat, whereas psychotherapists may ignore the physical behavior in favor of the verbal, the author argues for a combination of physical and psychological treatment for neuroses.

115   Brace, David K. *Measuring Motor Ability: A Scale of Motor Ability Tests.* New York: A. S. Barnes and Co., 1927. 138 pp. *P T*

Tests comprised of thirty different stunts were given to subjects, ages eight to forty-eight, to assess their agility, balance, control, strength, etc. Each item is described and illustrated, and procedures for administering and scoring these are given. Test validity, reliability, and scaling are assessed, together with low or negligible correlations with motor achievement tests and age, weight, and intelligence. Applications of the motor-ability scores are suggested.

116   Brackbill, Yvonne. "Extinction of the Smiling Response in Infants as a Function of Reinforcement Schedule." *Child Development* 29 (1958):115-124. *G T*

The smiling of four infants was reinforced intermittently by the behavior of the experimenter, who smiled and cuddled the child over a short period of time in which other social contacts were controlled. Four other babies were given regular reinforcement. In the "extinction" phase, the intermittently reinforced babies persisted in smiling longer than the regularly reinforced babies.

117   Brengelmann, J.C. "Expressive Movement and Abnormal Behaviour." In *Handbook of Abnormal Psychology,* edited by H. J. Eysenck, pp. 62-107. New York: Basic Books, 1961. *D G T*

The author focuses on objective measurement and controlled experimentation of expressive movement in relation to personality and psychiatric diagnosis. Many of the studies reviewed here deal with measurement of voice quality and artistic drawings and evaluation of "semiobjective" techniques such as the Bender-Gestalt Test, object-arranging tests, and handwriting analyses. Because these are largely perceptual-motor, research using them is not included in the present bibliography, and the reader is referred to Brengelmann's review and extensive bibliography for references to them. However, the section on "objective tests" of writing pressure, tracing, and "kinesthetic drawing" considered as experimental study of movement style and personality are notable here. For example, simple tracing of lines or drawings done while a subject is wearing lenses or is blindfolded may indicate rigidity or extroversion and may distinguish between psychotics and non-psychotics.

118   Breuer, Josef, and Freud, Sigmund. *Studies on Hysteria.* Translated by J. Strachey. New York: Basic Books, 1957. 335 pp.

Psychoanalytic interpretation of the traumatic origin or psychological significance of motor activity and muscular contractions can be found scattered throughout this book in the numerous references to hysterical symptoms, many of them motor, and in specific case histories such as that of Anna O., whose symptoms included paralysis or muscular contraction of limbs and neck (pp. 21-38); Frau von N., who had tics (pp. 49, 91-92) and neck cramps that recurred throughout her illness (pp. 70-71, 96-97); and an eighteen-year-old girl whose toe wriggling was shown to have specific psychic origin (pp. 93-95). Motor activity as discharge of excitation is also discussed (pp. 201-202).

119   Brewer, W. D. "Patterns of Gesture Among the Levantine Arabs." *American Anthropologist* 53 (1951): 232-237.

A description of Levantine Arab gestures considered by the author to have well-defined, almost "lexical" meaning is presented in three groups: (1) those with "symbolic" meaning, independent of speech; (2) those with "pictorial" meaning, often dependent on speech; and (3) those with "emphatic" meaning, comprehensible only with speech. Sixteen distinct gestures are described, together with a brief interpretation of their meanings and when they occur.

120   **Brody, Sylvia.** *Patterns of Mothering: Maternal Influence during Infancy.* New York: International Universities Press, 1956. 446 pp. *T* [Latest Edition: 1970]

The personalities, behavioral styles, and interaction patterns of thirty-two mother-infant pairs were studied in depth from live observation, film, psychological tests, interviews, etc., and particular focus was put on the feeding behavior as being most representative of the mother's relationship to her child. The behavioral observations are presented in detail, along with quantitative analyses of the material and psychoanalytic discussion of its significance. This book also contains a valuable review of mother-infant literature.

121   **Brosin, Henry W.; Condon, William S.; and Ogston, William D.** "Film Recording of Normal and Pathological Behavior." In *Hope: Psychiatry's Commitment,* edited by A. W. R. Sipe, pp. 137-150. New York: Brunner-Mazel, 1970.

A discussion of the slow evolution of linguistic-kinesic research and its value and potential to psychiatry; also, a survey of the specific research of Condon and Ogston. Almost every formulation and finding of these linguistic-kinesic researchers is cited. New material is included, such as an analysis of "dominance-submission" behavior in males.

122   **Brown, William H.** "An Investigation of the Relationship between Idiopathic Epilepsy and Peripheral Motor Activity." Ph.D. dissertation, University of Kansas, 1951. 127 pp. (Datrix order no. 5404)

A group of institutionalized epileptic patients were observed to perform simple motor tasks such as rhythmic tapping as well as more intelligent normal subjects, but they did not do so well in rhythmic coordination of hands and feet or with writing tasks.

123   **Bruch, Hilde, and Thum, Lawrence C.** "Maladie Des Tics and Maternal Psychosis." *Journal of Nervous and Mental Disease* 146 (1968): 446-456.

Case histories of a mother and her twelve-year-old son who developed severe tics which abated when his mother became psychotic. The etiology of the tic syndrome in relation to the family dynamics, particularly the mother's severe restraints on her son, is discussed.

124   **Bruner, Jerome S., and Taguiri, Renato.** "The Perception of People." In *Handbook of Social Psychology,* vol. 2, edited by G. Lindzey, pp. 634-654. Reading, Mass.: Addison-Wesley Publishing Company, 1954. [Latest edition: ed. by G. Lindzey and E. Aronson, 1968]

A classic critique of research on judgment of emotion from facial expression, judgment of personality from "external signs," and the nature of impression formation.

125   **Bühler, Charlotte.** "The Social Behavior of Children." In *A Handbook of Child Psychology,* 2d rev. ed., edited by C. Murchison, pp. 374-416. Worcester, Mass.: Clark University Press, 1933. *G T* [New York: Russell & Russell Publishers, 1967]

An excellent review of early research on the development of social behavior from early infancy to adolescence, with numerous references to facial expression, gesture, and nonverbal communication. It summarizes studies (mostly German or American) on the onset of smiling and what evokes it; how the infant responds to "angry" versus "kind" body expression differently at different ages; the various stages and kinds of social contact between babies and between baby and adult; when the child is able to relate to more than one person; subgrouping patterns in young children; the development of leadership and what behaviors reflect it from six months on; patterns of group formation; and the effects of different environments on the child's social contacts.

126   **Bull, Nina.** *The Attitude Theory of Emotion.* Nervous and Mental Disease Monographs, no. 81, 1951 (reprint ed., New York: Johnson Reprint Corporation, 1968). 159 pp. *D T*

Bull's attitude theory states that there are five stages leading to action: stimulus; predisposition, or latent attitude; motor attitudes that are both preparatory and communicative in nature; feelings or mental attitudes resulting from the motor postures of readiness; and finally, the consummatory action. Other theories of the origin of emotion and thought (e.g., James-Lange, Freud, Cannon) are discussed relative to this one. Then research on hypnotized subjects, in which suggestions that the subject would feel certain emotions led to characteristic postures, is reported to support the theory. Motor patterns elicited for fear, disgust, anger, depression, and joy are described along with the subject's introspective reports. There is also a chapter on emotion and patterns of visual fixation.

127   **Bunzel, Gertrude G.** "Psychokinetics and Dance Therapy." *Journal of Health and Physical Education* 19 (1948):180-181, 227-229. *P*

An introduction to the author's "psychokinetic method" of dance therapy with normal children and adults.

128   **Burton, Arthur, and Kantor, Robert E.** "The Touching of the Body." *Psychoanalytic Review* 51 (1964):122-134.

A discussion of therapists' avoidance of touching their patients as a counter-transference problem; the cultural and psychological significance of touch; and the need for therapist and patient to feel free to touch each other.

129   **Byers, Paul, and Byers, Happie.** "Nonverbal Communication and the Education of Children." In *Functions of Language in the Classroom,* ed. by C. Cazden, V. John, and D. Hymes. New York: Columbia University Teachers College Press, forthcoming.

The authors begin with a discussion of the nature of human communication and the role of nonverbal communication in interpersonal relatedness. They contrast the new view of face-to-face communication as complex, multi-leveled message-sending with traditional over-emphasis on language and speech content. Citing specific examples of a young child learning to communicate and encounters between teacher and schoolchild, the authors stress that the focus must be on what occurs between people and the context in which the interaction occurs. Cultural differences in patterns of eye-contact, touch and timing are illustrated by descriptions of interactions between white teacher and Puerto Rican and Black children. The significance of cultural differences in codes of nonverbal communication is discussed in relation to education.

130   **Canna, D. J., and Loring, Eugene.** *Kineseography: The Loring System of Dance Notation.* N.p.: Academy Press, 1955. 57 pp. *D*

A comprehensive movement notation developed from dance which employs a staff, arrows for directions, symbols for body postures, and the movements of parts in terms of "extrovert" (outward), "normal," and "introvert."

131   **Canner, Norma.** *. . . And a Time to Dance.* Photographed by H. Klebanoff. Boston: Beacon Press, 1968. *P*

In a sensitive work replete with photographs of children and teacher moving together, the author discusses how movement may be used therapeutically with retarded children, what may be done in sessions, and how body awareness, communication, and concepts are developed through movement.

132 **Cannon, Walter B.** *Bodily Changes in Pain, Hunger, Fear and Rage.* New York: D. Appleton & Company, 1929. 404 pp. *D G* [College Park, Md.: Mc-Grath Publishing Co.]

Although most of this classic work deals with physiological changes accompanying intense needs and emotions, there are chapters on reactions of fatigued muscles and the "energizing effect" of the emotions on body movement, the physiological function of motor attitudes in extreme emotion, mania, rituals, and dancing and fighting, as well as scattered observations of typical postural reactions and body movements under strong emotion, following removal of animal cortex, and so on.

133 **Carmichael L.; Roberts, S. O.; and Wessell, N. Y.** "A Study of the Judgement of Manual Experiences as Presented in Still and Motion Pictures." *Journal of Social Psychology* 8 (1937):115-142. *P T*

Subjects judged the emotion or expression conveyed by thirty-five photographs of an actor's hands in different positions (the photographs are reprinted), and there was high agreement. Making judgments of films of only the hands in thirty-five gestures, a different group showed still higher agreement. Differential judgment between "conventionalized" and "emotional" gestures was not found.

134 **Carpenter, C. R.** "A Field Study of the Behavior and Social Relations of Howling Monkeys." *Comparative Psychology Monographs, Serial no. 48, 10* (1934):1-168. *D P T*

An impressive early field study, this includes descriptions of the monkeys' reactions to observers, their manual dexterity, their postures and locomotion patterns, mother-infant behaviors, subgrouping and interaction patterns, male-female differences and sexual behavior, relationships among males, and movement and vocal behaviors that coordinate interaction.

135 **Carpenter, C. R.** "A Field Study in Siam of the Behavior and Social Relations of the Gibbon (Hylobates Lar)." *Comparative Psychology Monographs,* Serial no. 84, 16 (1940):1-212. *P T*

Similar in format to Carpenter's previous field report, this includes discussion of the animals' reactions to being observed, their postures and stance, locomotion, prehension, excited behavior, sexual activity, minor male-female differences in social behavior, mother-father-infant interaction, play activity, greeting behaviors, actions affecting group coordination, and dominance and aggressive behaviors.

136 **Carpenter, C. R.** "Sexual Behavior of Free Ranging Rhesus Monkeys. I, Specimens, Procedures and Behavioral Characteristics of Estrus." *Journal of Comparative Psychology* 33 (1942):113-142. *T*

A concise description of the sexual behaviors and physiological changes of rhesus monkeys during estrus, including modes of presenting postures, patterns of attack and submission, and copulation behaviors. Also included is a good reference list of studies of sexual behavior in other primates.

137 **Carpenter, C. R.** *Naturalistic Behavior of Nonhuman Primates.* University Park: The Pennsylvania State University Press, 1964. 454 pp. *D G P T*

A collection of Carpenter's papers representing nearly thirty years of field research and analysis of the behavior of monkeys and apes. Motor patterns, signal behaviors, and body expression are dealt with in discussions of dominance patterns, territorial behavior, group interaction and regulation, sexual activity, mother-infant interaction. and so on.

138   **Carrington, W. H. M** "The F. Matthias Alexander Technique: A Means of Understanding Man." *Systematics* 1 (1963):233-247.

A clear exposition of Alexander's theories about and techniques for effecting change of physical-mental habits perceived in body carriage.

139   **Casa, Giovanni della.** *Galateo of Manners and Behaviours.* Boston: Merrymount Press, 1914, 123 pp.

Written in the mid-sixteenth century, this treatise contains numerous references to how Renaissance Italian gentlemen or ladies should walk, express themselves, and behave.

140   **Cason, Hulsey.** "Common Annoyances: A Psychological Study of Everyday Aversions and Irritations." *Psychological Monographs,* Whole No. 182, 40 (1930):1-218. *G*

An ambitious survey of what Americans of that time considered annoying, including 1,523 different types of behavior. It is of interest here as a source of information about judgments of certain nonverbal behaviors.

141   **Cason, Hulsey.** "The Influence of Tension and Relaxation on the Affectivities." *Journal of General Psychology* 18 (1938):77-110. *T*

Subjects were asked to report what feelings were evoked in them by different photographs, under conditions of muscle tension (pressing hard on a pedal), relaxation (semirecumbent), and normal sitting. The subject also pressed a key "as long as any emotion was present." In a number of ways, tension or relaxation had no clear effect on their reports, although tension increased the latent time of response and relaxation affected the duration of the emotion.

142   **Causley, Marguerite.** *An Introduction to Benesh Movement Notation: Its General Principles and Its Use in Physical Education.* London: Max Parrish and Company, 1967. 91 pp. *D N*

A presentation of the widely used Benesh notation which shows in detail how rhythm, dynamics, locomotion, movements of head and trunk, and movements in kneeling, sitting, and lying down are recorded.

143   **Chace, Marian.** "Dance as an Adjunctive Therapy with Hospitalized Mental Patients." *Bulletin of the Menninger Clinic* 17 (1953):219-225.

In this paper the late Marian Chace, for many years the foremost American dance therapist, describes the setting, structure, and techniques of dance therapy in her work with severely disturbed patients at St. Elizabeth's Hospital and Chestnut Lodge Sanitarium. Ways of initiating contact, responding to individual needs, and developing group interaction with various types of patients are described. In her interpretation of specific body movements and postures, her description of how she responds, and her analysis of the phases of a dance therapy session, Miss Chace presents the principles behind her use of dance and music as a primary means of reintegrating the seriously ill mental patient into a group.

144   **Chace, Marian.** "Use of Dance Action in a Group Setting." Presented at the American Psychiatric Association Convention, Los Angeles, California, 1953. 8 pp. (Copies from Dance Therapy Section, St. Elizabeth's Hospital, Washington, D.C.)

Movement characteristics of three types of hospitalized psychiatric patients—manic, depressed, and schizophrenic—are described, particularly as observed in the patient's dancing. Examples of how Marian Chace relates to individual patients through dance are presented. She describes how the dance therapist "answers" the patient with similar movement to establish contact, to encourage the patient's expression of his conflict in movement, and ultimately to facilitate his involvement in the group.

145   **Chace, Marian.** "Dance Therapy for Adults." Presented at the National Education Association Conference, Atlantic City, n.d. 9 pp. (Copies from Dance Therapy Section, St. Elizabeth's Hospital, Washington, D.C.)

A theoretical paper about the function of dance therapy in a hospital setting, the role of the dance therapist, and the value of dance for helping psychiatric patients vent their feelings and become less isolated.

146   **Chance, M. R. A.** "An Interpretation of Some Agonistic Postures: The Role of 'Cut-off' Acts and Postures." *Symposium of the Zoological Society of London* 8 (1962):71-89. *D*

An analysis of protective movements or opposition in certain rats and birds when they are threatened by, but attracted to, another animal—in terms of specific postures, eye closing, and/or turning away of the head, thereby "cutting off" the social stimulus and decreasing arousal. Also, a discussion of different courtship behaviors in birds and the function of "cut-off" behaviors in controlling arousal and preventing flight.

147   **Chaney, Clara M., and Kephart, Newell C.** *Motoric Aids to Perceptual Training.* Columbus, Ohio: Charles E. Merrill Publishing Co., 1968. 138 pp. *P*

A presentation of motor and perceptual excercises for training brain-injured and retarded children: theoretical rationale, evaluation (includes the Purdue Perceptual-Motor Survey), and description of the activities and what they are designed to develop (balance, locomotion, differentiation, eye coordination, body image, etc.).

148   **Chapple, Eliot D., with the collaboration of Conrad M. Arensberg.** "Measuring Human Relations: An Introduction to the Study of the Interaction of Individuals." *Genetic Psychology Monographs* 22 (1940):3-147. *D T*

Although Chapple's method of interaction analysis does not explicitly record what kind of movements occurred, his pioneering work—introduced in this monograph—helped pave the way for later communication research that attended to the kinesic level. Special-usage terms ("action," "event," "component," and "set"); equipment for recording interaction sequences; analysis of interaction patterns in various settings; and interaction hierarchies, systems, and subsystems are defined and/or assessed.

149   **Charny, E. Joseph.** "Psychosomatic Manifestations of Rapport in Psychotherapy." *Psychosomatic Medicine* 28 (1966):305-315. *G T*

Further developing A. Scheflen's findings on the function of body positions in psychotherapy, Charny presents evidence that specific postural configurations indicate relatedness between patient and therapist. "Congruent" and "noncongruent" postures of a male therapist and female patient were observed, frame by frame, in a 33-minute film and compared with a structural or semological analysis of the speech. Charny found that "upper-body mirror congruent posture" occurred together with speech reflecting greater relatedness (i.e., more positive, specific, and addressed to the therapy situation); whereas noncongruent positions occurred when the speech was self-centered, negational, or self-contradictory.

150   Checkov, Michael. *To the Actor: On the Technique of Acting.* New York: Harper & Brothers, 1953. 201 pp. *D*

Movement is integral to this director's methods of actor training, and this book is a valuable example of how an actor-director understands body movement. He presents exercises designed to develop the actor's movement, explore expression of feeling, and enhance improvisation and group interaction. The chapter entitled "The Psychological Gesture" explores the significance of various movements, the relation between performing a movement and experiencing new feelings, and the development of a character from a significant core movement or gesture.

151   Cho, Won-Kyung. "Dances of Korea." Unpublished paper. 38 pp. *D N P* (Copies from Dance Notation Bureau, 8 East 12th St., New York)

The author, an accomplished Korean classical dancer, sketches a history of Korean dance and culture and outlines the religious, court, and folk dances in terms of who performs them, when, and with what costumes and musical accompaniment. There are brief descriptions of the dances and their meaning, illustrations of dance positions, and a Labanotation score for one court dance.

152   Christiansen, Bjorn. *Thus Speaks the Body: Attempts toward a Personology from the Point of View of Respiration and Postures.* Oslo, Norway: Institute for Social Research, 1963. 235 pp. *D T*

The scope of this valuable yet hard-to-obtain monograph is enormous. As Christiansen reviews literature on muscular tension, postures, gestures, respiration patterns, and body image in relation to personality and psychodynamics, he summarizes findings of a wide spectrum of writers, from Reich to Malmo, and elaborates on them with his own formulations. He begins with an excellent summary of the work of Mahl and Deutsch and their specific hypotheses about which body parts and movements appear to correlate with which psychodynamic themes or conflicts. He goes on to review literature on the relation of breathing patterns to personality, mental states, and sex differences and to discuss the physiology of breathing in erect and supine postures. Included is an extensive review of work on differences in breathing patterns of normal, neurotic, and schizophrenic individuals, breathing patterns of anxiety states and specific neuroses, and the effects of postural changes on respiration. Following a section on breathing characteristics of infants and children, he explores developmental aspects of basic organismic and emotional responses (e.g., sucking, laughing, orgasm), breathing and posture conceptualized within the framework of psychosexual stages, and "phase modalities" or "generalized modes of reacting." In this he integrates work of Freud, Reich, Lowen, Spitz, and Braatöy and takes issue with Erikson. Pursuing the notion that personality and psychological conflicts are revealed in posture (here "posture" most often means characteristic stance, positions, and patterns of muscular tension in the body), he discusses EMG studies of muscular tension as indicators of conflict, "nervous overflow," "peripheral inhibition," and situational change. Asymmetry, or right-left differences, and generalized versus localized differences in tension are focused upon. He concludes with a synthesis of the literature on posture and interpersonal perception, postural tension and body awareness, body image as expressed in figure drawings, tension and pain, body-image boundary and muscular tone, and muscular tension and vasomotor conditions. This monograph has an excellent bibliography, with references to articles on breathing patterns in relation to personality, development, affect, and cognition.

153   Christrup, Helen J. "The Effect of Dance Therapy on the Concept of Body Image." *Psychiatric Quarterly Supplement, Part 2,* 36 (1962):296-303. *T*

An experiment on the effectiveness of dance therapy with groups of chronic schizophrenic patients, as measured by changes in the Goodenough scores and Swenson Sexual Differentiation scores on the patients' projective drawings done before and after thirteen weeks of dance therapy sessions. In this well-executed study, dance therapy was the only formal therapy other than drugs used for the experimental groups. Patients who were most active in sessions as determined by two raters showed significant improvement; and compared with the control groups, the female patients who had dance therapy showed significant improvement on the Goodenough scores.

154 **Clarke, Frances M.** "A Developmental Study of the Bodily Reaction of Infants to an Auditory Startle Stimulus." *Journal of Genetic Psychology* 55 (1939): 415-427. *T*

The startle response of fourteen infants, filmed at seven intervals up to age twenty weeks, is minutely described.

155 **Cline, Marvin G.** "The Influence of Social Context on the Perception of Faces." *Journal of Personality* 25 (1956):142-158. *D T*

Subjects looked at schematic drawings of two faces "looking at" each other with different expressions (e.g., one smiling, the other frowning), described the situation, and rated them according to the degree to which a figure was dominant, initiating, etc. Subjects consistently perceived interaction, and this influenced their judgments of the expressions.

156 **Clynes, Manfred.** "On Being in Order." *Zygon: Journal of Religion and Science* 5 (1970):63-84. *D G*

Within a discussion which is sometimes metaphysical and sometimes poetic, Clynes describes some innovations in the study of emotion. Fine measures of finger pressure on a key and muscle potentials of arm and back are made as the subject concentrates on an emotion such as anger, love, or grief. The research indicates that there are characteristic patterns of horizontal and vertical movement and muscle activity produced for each emotion which is fantasized. He discusses the origins, neurophysiological aspects, and implications of these patterns, which he calls "essentic forms." He proposes that an essentic form may be expressed in analogous ways through movement, music, tone of speech, and art. The article includes some fascinating remarks on essentic forms, empathy, and communication processes and on how performing a series of pure essentic forms gives one a feeling of well-being. Clynes is both a noted concert pianist and chief research scientist at the Biocybernetics Laboratory, Rockland State Hospital Research Center, Orangeburg, N.Y. Other writing by him—much of which appears to be in technical computer and medical terminology—is cited in this paper.

157 **Coleman, James C.** "Facial Expressions of Emotion." *Psychological Monographs*, Whole No. 296, 63 (1949):1-36. *P T*

This has an excellent, concise review of research on judgment of facial expression. In the study reported, several subjects who went through a series of often-distressing stimuli (e.g., shock, crushing a snail) were filmed and later were asked to reenact each expression. Observers judged films of either a male subject, natural and acting, or a female subject, natural and acting, with either upper face, lower face, or entire face showing, and they selected the description that matched the film segment. In the filmed experimental part, subjects' reactions were varied both in expression and in their report as to what they felt, and individual differences were more marked than any sex differences. In the judgment part of the study, certain expressions were better judged from the upper face, others from the lower; there were no sex differences in judgment; laughter was most easily identified, reaction to threatened shock hardest; and in some cases whether the expression was acted or natural affected the judgment.

158   Coleman, Roy; Greenblatt, Milton; and Solomon, Harry C. "Physiological Evidence of Rapport During Psychotherapeutic Interviews." *Diseases of the Nervous System* 17 (1956):71-77. G

While this study primarily measures the heart rates of a patient and a therapist during psychotherapy sessions, there are clinical judgments of emotions based on voice quality and types of body movement which are correlated with the heart rates. It provides evidence of which heart rates are characteristic of which emotions (and, by inference, which body movement characteristics).

159   Committee on Research in Dance. *Workshop in Dance Therapy: Its Research Potentials.* Proceedings of a Joint Conference by the Research Department of Postgraduate Center, Committee on Research in Dance, American Dance Therapy Association, 1970. 69 pp.

In the conference reported here a number of dance therapists, psychologists, psychiatrists, and anthropologists met to discuss research in dance therapy, using videotapes of three different approaches to the psychotherapeutic use of body movement as reference points. What emerges is a lively discussion about the nature of dance therapy and the interpretation of movement patterns and a heated debate about the intrapsychic versus the cultural basis of movement patterns.

160   Condon, Williams S. "Lexical-Kinesic Correlation." Presented at Stagecoach Meeting, n.p., February 1963. 6 pp. *T* (Copies from author, School of Medicine, Western Psychiatric Institute and Clinic, 3811 O'Hara St., Pittsburgh, Pa. 15213)

Following repeated film viewing of a woman in a psychiatric interview, tabulation was made of her recurring gestures and the verbal contexts in which they occurred. The interview appeared to have two phases, marked by shifts in the verbal content and in the number of certain gestures.

161   Condon, William S. "Synchrony Units and the Communicational Hierarchy." Unpublished paper, 1963. 16 pp. *D* (Copies from author, School of Medicine, Western Psychiatric Institute and Clinic, 3811 O'Hara St., Pittsburgh, Pa. 15213)

In this unpublished paper Condon introduces his early findings on interactional synchrony, in which changes in the direction or speed of the body movement of two or more people occur simultaneously. Using a natural history "search/check" method for defining units of communication from film, he proposes that synchrony units may constitute small bits at one level in a communication hierarchy, head directional units the next, arm directional units the next, and what he calls "holding units" the next. He concludes with a discussion of the possible uses and ramifications of these formulations.

162   Condon, William S. "Process in Communication." Unpublished paper, 1964. 17 pp. *D* (Copies from author, School of Medicine, Western Psychiatric Institute and Clinic, 3811 O'Hara St., Pittsburgh, Pa. 15213)

A further examination of synchrony with better-quality film, at 48 frames per second. Condon continues to explore and delineate the hierarchy of behavioral units and which types of kinesic changes correspond to which linguistic changes at increasing levels of organization. In what he calls level four, for example, the beginning to end of the direction of a head and trunk movement corresponds to the beginning and end of a word. Six levels of units are postulated, with an additional discussion of self-synchrony and interactional synchrony. Condon cites an incident in which a child was observed trying to move synchronously with his "dyssynchronously" moving parents and in the process began to move bizarrely. The author's observations are so compelling and important that one wishes for greater corroboration, particularly for establishing observer agreement.

163 **Condon, William S.** "Linguistic-Kinesic Research and Dance Therapy." *American Dance Therapy Association Proceedings, Third Annual Conference*, pp. 21-44, 1968. *D* (Copies from ADTA, 5173 Phantom Court, Columbia, Md. 21043)

A theoretical discussion of order and rhythm in behavior and problems of defining units in the continuous flow of behavior, together with a summary of Condon's observations of "self-synchrony" (changes in direction correspond to changes in speech articulation), "interactional synchrony" (motion of the listener is synchronous with the speech and motion of the speaker), and the hierarchical organization of linguistic-kinesic units. Condon cites research showing "self-dyssynchrony" among individuals with aphasia seizures, parkinsonism, stuttering, or schizophrenia and focuses on film analysis of a three-and-a-half-year-old girl diagnosed as autistic who would move to the diverse sounds in the room, thus appearing bizarre and even deaf. Further, a minute film analysis of a mother and her twins, one of whom was schizophrenic, showed that the mother moved synchronously with her "well" daughter, but when the schizophrenic girl adopted a posture like her mother's, the mother would quickly change. Condon suggests that dance therapy may be one way to reestablish synchrony in behavior, and hence to "re-initiate communication."

164 **Condon, William S.** "Method of Micro-analysis of Sound Films of Behavior." *Behavior Research Methods and Instrumentation* 2 (1970):51-54. *D*

A practical and concise presentation of how Condon analyzes sound films frame by frame for the synchrony and hierarchic organization of speech and motion patterns. He describes the equipment used, what modifications he makes on it, and how he operates it. The analysis of a 23-frame film segment is used to illustrate how he observes and records changes in direction and body part and how the movement patterns correspond to hierarchic levels of syntactic organization. He then briefly reviews his and his colleagues' research on interactional synchrony.

165 **Condon, William S., and Brosin, Henry W.** "Micro Linguistic-Kinesic Events in Schizophrenic Behavior." In *Schizophrenia: Current Concepts and Research*, edited by D. V. Siva Sankar, pp. 812-837. Hicksville, N.Y.: PJD Publications, 1969. *D T*

First there is a discussion of normal "self-synchrony" and "interactional synchrony" as researched through very detailed analysis of film at 24 or 48 frames per second. The authors then discuss and illustrate "dyssynchrony" in the behavior of a stutterer, a chronic schizophrenic, a woman with multiple personality, and an autistic child. They then focus on the nature of the self-dyssynchrony in schizophrenia, particularly the various ways that the body parts are "out of phase" with each other in three different patients. The authors speculate about the interpersonal significance of this dyssynchrony and how it differs from that seen in organic disabilities.

166 **Condon, W. S., and Ogston, W. D.** "Sound Film Analysis of Normal and Pathological Behavior Patterns." *Journal of Nervous and Mental Disease* 143 (1966):338-347 *D*

The authors present an example of synchrony between the articulated speech and the movement of both speaker and hearer and note that they have observed this phenomenon in about thirty films of normals. They report that they have observed marked dyssynchrony within the speech and body motion of an aphasic patient. Similarly a minute analysis of a schizophrenic patient's speech and movement indicated self-dyssynchrony and lack of variation and rhythmic mobility. Specific abnormal features of the patient's motion and speech are listed, along with a sample linguistic-kinesic analysis of his behavior.

167   Condon, W. S., and Ogston, W. D. "A Method of Studying Animal Behavior." *Journal of Auditory Research* 7 (1967):359-365. D

Condon and Ogston apply the methods of linguistic-kinesic research with which they examined synchrony of motion and speech in humans to the interaction between a man and a chimpanzee he has trained. They present diagrams showing that the chimp and the man are in synchrony while he speaks and, further, that the movement of the chimp is in "self-synchrony" with its own vocalizations in a pattern similar to humans.

168   Condon, W. S., and Ogston, W. D. "A Segmentation of Behavior." *Journal of Psychiatric Research* 5 (1967):221-235. D T

Discusses the application of linguistic "search/check" methods of defining behavior units and patterns to the study of linguistic-kinesic patterns in communication. From analysis of film down to 48 frames per second, examples are presented of "self-synchrony" (i.e., the movement of a speaker largely in terms of changes in direction is synchronous with changes in his speech) and "interactional synchrony" (in this case, a father and son move synchronously with the speech and motion of the mother). In addition, the authors present data indicating that polygraph pen changes in EEG recordings are synchronous with changes in speech and motion.

169   Condon, W. S., and Ogston, W. D. "Speech and Body Motion Synchrony of the Speaker-Hearer." In *The Perception of Language*, by D. L. Horton and J. J. Jenkins, pp. 224-256. Columbus, Ohio: Charles E. Merrill Publishing Co., 1971. D

A theoretical discussion of the problems of defining units and patterns of behavior, with a presentation of the authors' findings on self-synchrony and interactional synchrony. Primarily, however, this paper is a report of research on the relationship of eye blinking to speech. The authors state that of the 718 eye blinks they studied on film, most occur exactly at the beginning of a word or at an "articulatory change point" within the word.

170   Condon, William S.; Ogston, William D.; and Pacoe, Larry V. "Three Faces of Eve Revisited: A Study of Transient Microstrabismus." *Journal of Abnormal Psychology* 74 (1969):618-620. T

An actual film of the woman with multiple personality portrayed in the book and film *Three Faces of Eve* was minutely examined frame by frame. The authors detected an eye strabismus occurring for 1/8 of a second or less, and thus only observable in slow motion or stop-frame analysis. The rate and kind of strabismus varied with the different personalities of Eve, and no eye divergence was observed in the patient later in therapy when she was considered most improved. The authors state that they have seen this disturbance in oculomotor parallelism in films of schizophrenic patients but not in films of normal subjects.

171   Copple, Lee Biggerstaff. "Motor Development and Self-Concept as Corre-

lates of Reading Achievement." Ph.D. dissertation, Vanderbilt University, 1961. 52 pp. (Datrix order no. 61-3596)

Using a study sample of 102 fifth-grade boys, the author found almost no correlation between reading ability and motor development or athletic ability (as measured by tasks involving speed, hand-eye coordination, muscular power, etc.). Only one motor task, involving left-handed grip, correlated positively with reading achievement.

172　Corbin, Edwin I. "Muscle Action as Nonverbal and Preverbal Communication." *Psychoanalytic Quarterly* 31 (1962):351-363.

From the psychoanalysis of a woman who had distinctive muscular tensions and motor habits, the author interprets the function of motor action in regression and displacement of aggressive impulses.

173　Corsini, Raymond J. *Roleplaying in Psychotherapy: A Manual.* Chicago: Aldine Publishing Company, 1966. 206 pp.

Only a few of the examples described in this book on theoretical and methodological aspects of psychodrama include analyses of movement or body expression. However, it has an excellent 133-item annotated bibliography on psychodrama literature, by Samuel Cardone.

174　Cranach, Mario von. "The Role of Orienting Behavior in Human Interaction." In *Behavior and Environment: The Use of Space by Animals and Men,* edited by A. H. Esser, pp. 217-237. New York: Plenum Publishing Corporation, 1971.

An excellent review of research on eye contact, gaze behavior, and eye-head orienting responses, with a bibliography that includes a number of German papers on the subject. Von Cranach's discussion has sections on the definition of various types of looking, the reliability of observer assessment of gaze and eye contact, factors influencing observer assessment, eye region versus gaze as releasers of the infant's smile, relationships between distance and looking behavior, the effects of looking on the receiver, communicational significance of gaze avoidance, gaze and the regulation of interaction, and eye and head movements in greetings.

175　Cranach, Mario von, and Vine, Ian, eds. *Expressive Movement and Nonverbal Communication.* London: Academy Press, forthcoming.

176　Crane, George E., and Paulson, George. "Involuntary Movements in a Sample of Chronic Mental Patients and Their Relation to the Treatment with Neuroleptics." *International Journal of Neuropsychiatry* 3 (1967):286-291. *T*

A survey of 182 chronic mental patients which shows that 15 percent had "tardive dyskinesia" (choreiform of athetoid movements, such as jerky, irregular movements of extremities, etc.) and that 7 percent had parkinsonism-like symptoms. The authors discuss tardive dyskinesia as an apparent result of long-term phenothiazine treatment.

177　Cratty, Bryant J. *Movement Behavior and Motor Learning.* 2d rev. ed. Philadelphia: Lea & Febiger, 1967. 367 pp. *D G*

As a physical education textbook, this work focuses on motor performance; there are also chapters on body movement characteristics in relation to personality and ability traits and on the effects of anxiety and stress on motor performance. It is particularly valuable for its review of the literature and a 1,000-

title bibliography on motor skills and learning, perceptual aspects of motor performance, and developmental aspects of movement.

178 Cratty, Bryant J. *Psychology and Physical Activity.* Englewood Cliffs, N.J.:
Prentice-Hall, 1968. 214 pp. *D*

Focusing on psychological factors in motor performance and sports, this book
has an unusual format. It is a series of hypotheses culled from research and their
implications for physical education. As such it summarizes a considerable amount
of research on personality and motor performance; motivational, developmental,
perceptual, and learning aspects of motor skill; and the effects of an audience,
group interaction, and competition on motor performance, with additional observations on superior athletes and "clumsy" children. In effect, the book goes
beyond its field to become a rich source of hypotheses for research.

179 Cratty, Bryant J. *Perceptual and Motor Development in Infants and Children.* New York: The Macmillan Company, 1970. 306 pp. *D G T*

A valuable bibliographic source and survey of research on motor development.
The following topics are explored: reflexes and development of locomotion in the
infant; laterality, locomotion patterns, and gross motor skills in young children;
development of visual perception and body image; development of prehension,
drawing, and writing; the relationship of physical skills to intelligence; evaluation of motor skills in older children; analysis of "perceptual-motor programs,"
sex differences, and social factors in motor development.

180 Critchley, Macdonald. *The Language of Gesture.* London: Edward Arnold,
1939. 128 pp. *T* [New York: Haskell House Publishers, 1970]

An historic work in its own right, this book is a valuable source of historical
references to movement and gesture, particularly Roman and Greek writing on
gesture in oratory, mime, and dance (noting Plutarch, Lucian, Quintilian, Ci-
cero). There are chapters on the gesture origins of speech; neurophysiological aspects of gesture; the history and nature of deaf sign language; the sign language
of American Indians, Australian Aborigines, religious orders, and secret societies;
the symbolic gestures and postures of Hindu and Chinese theater; universal aspects of sign language; and characteristics of Arab, Jewish, and Italian gestures
(with special note of Italian literature on gestures).

181 Critchley, Macdonald. "Kinesics: Gestural and Mimic Language, An Aspect
of Non-verbal Communication." In *Problems of Dynamic Neurology*, edited by
L. Halpern, pp. 181-200. Jerusalem: Hebrew University, Hadassah Medical
School, 1963. *P* [New York: Grune & Stratton]

Critchley surveys many types of gestures and discusses the role of gesture and
body expression in speech and communication; cultural differences in gesture and
body movement characteristics; and universal gesture symbols. He cites some
rare sources on deaf sign language and gestures in Roman oratory, and notes
types of gestures of workmen, Indian dancers, and other groups.

182 Crown, Sidney. "An Experimental Inquiry into Some Aspects of the Motor
Behaviour and Personality of Tiquers." *Mental Science* 99 (1953):84-91. *T*

Nine patients with tic and nine controls were given tests of movement skill, "expressive movement," and personality, (e.g., Luria test, manual dexterity). Tiquers were consistently better in skilled movement tests, and no differences were
found in expressive movement tests; however, the voluntary movement of tiquers
was more easily disorganized under the influence of emotion.

183 **Cuceloglu, Dogan M.** "A Cross-Cultural Study of Communication via Facial Expressions." Ph.D. dissertation, University of Illinois, 1967. 228 pp. (Datrix order no. 68-8046)

Individuals from different cultures rated sixty abstract facial expressions on forty emotion scales. A factor analysis of the data showed that cross-culturally three dimensions (pleasantness, activity, and control) accounted for most of the variance, whereas a fourth factor (a "cognitive-gut" dimension) was unique to the Japanese interpretations of facial expression. Evidence supporting the notion of cross-cultural "model postures" of facial expression was obtained. Also, subjects showed high agreement on "affective meaning," but not on "referential" or "denotative" meanings.

184 **Cunningham, Bess V.** "An Experiment in Measuring Gross Motor Development of Infants and Young Children." *Journal of Educational Psychology* 18 (1927):458-464. *T*

Infants' performance on a series of structured motor tasks was compared with mental age and body build. The series of tests are listed here in order of difficulty, from ages twelve to thirty-six months.

185 **Cutner, Margot.** "On the Inclusion of Certain 'Body Experiments' in Analysis." *British Journal of Medical Psychology* 26 (1953):263-277.

Clinical descriptions of how this Jungian analyst incorporated "body experiments" such as relaxation and body awareness exercises at key points in therapy sessions and the ways in which these facilitated treatment are presented. Two patients, a manic-depressive woman and a man with severe migraines, are discussed at length, and incidents with other patients are described briefly. Relationships between the patients' symptoms, conflicts, dreams, and body carriage and tensions are extensively described and interpreted.

186 **Da Costa, Maria I. L.** "The Ozeretzky Tests: Method, Value, and Results (Portuguese Adaptation)." Translated by E. J. Fosa. *Training School Bulletin* 43-44 (1946):1-13, 27-38, 50-59, 62-74. *D*

The Ozeretzky Tests (sometimes spelled Oseretsky) of motor maturation were first published in 1923 in Russian. Although they are frequently cited, this is a rare English publication of the test itself: test items, procedures, equipment, and score sheet. The test is designed to assess static and dynamic coordination, motor speed, simultaneous voluntary movements, and associated involuntary movements. It is presented in four installments, according to tests for four, six, ten, and fifteen or sixteen years of age. The items include such tasks as standing on tiptoe, throwing a ball at a target, and alternately closing the right and the left eye.

187 **Dance Index.** 7 vols. (1942-1948; reprint ed., New York: Arno Press, 1971). approx. 1,700 pp. *D N P*

A republication of all issues of *Dance Index* from 1942 through 1948, with a cumulative index prepared for this permanent, seven-volume collection. First edited by Lincoln Kirstein, Paul Magriel, and Baird Hastings, *Dance Index* is a major source of information on historical, ethnic, and theatrical dance, notable for its scholarship and for an extraordinary collection of prints and photographs. In addition to articles about specific dancers and choreographers, dance in various cultures, and dance criticism, there are special issues such as one on the history of dance notation and another on Picasso's work for the ballet.

188 **Darwin, Charles.** *The Expression of the Emotions in Man and Animals.*

New York: Philosophical Library, 1955, 372 pp. *D P* [Chicago: University of Chicago Press, 1965]

Originally published in 1872, after a century this work has no equal as a treatise on facial and body expression. Darwin draws on physiological descriptions, animal observations, and the reports of colleagues on expression in the insane and in various peoples around the world. He posits three principles of expression: "serviceable associated habits"; antithesis of expression, in which one is "serviceable" but its opposite is not; and "direct action of the nervous system," or excess of "nerve-force," leading to expressive and physiological reactions. Darwin describes the expression of different sounds, erection of hair or feathers, inflating the body, drawing back of the ears, retraction of lips, body positions, and head and limb movements of animals, particularly of the dog, horse, and monkey. He then analyzes the facial and body expressions of emotions in man, such as suffering, grief, joy, tender feelings, reflection, anger, and fear, always with examples from his own observations, anthropological reports, etc. He concludes with a chapter summarizing his theories about the evolution, function, and innate nature of bodily expressions.

189   **Dashiel, J. F.** "A New Method of Measuring Reaction to Facial Expression of Emotion." *Psychology Bulletin* 24 (1927):174-175.

A brief summary of a procedure for assessing children's judgments of facial expression, which involves their picking from a series of pictures the one that matches a short story. This method avoids vocabulary problems, differentiates ages better, and shows that younger children are able to discriminate subtler emotions than previous studies had indicated.

190   **Davey, A. G., and Taylor, L. J.** "The Role of Eye-Contact in Inducing Conformity." *British Journal of Social and Clinical Psychology* 7 (1968):307-308.

A study of whether several group members affected a subject's conformity more by looking at him than if there were no eye contact; the results are negative.

191   **Davis, D. Russell.** "Disorders of Skill: An Experimental Approach to Some Problems of Neurosis." *Proceedings of the Royal Society of Medicine* 40 (1947): 583-584.

The destructive effects of anticipatory tension found in cockpit performance tests are summarized. The implications of these studies for a theory of neuroses are outlined.

192   **Davis, Flora.** *Beyond Words: The Science of Nonverbal Communication.* New York: McGraw-Hill Book Company, forthcoming.

A review of current research in body movement, proxemics, and the role of the senses in communication. The research of Birdwhistell, Scheflen, Hall, Kendon, Goffman, Exline, Ekman, Bartenieff, M. Davis, Byers, Condon, and Chapple is discussed. The fact that these researchers were personally interviewed and often discussed work in progress makes this book particularly up-to-date and interesting.

193   **Davis, Martha.** "An Effort-Shape Movement Analysis of a Family Therapy Session." Unpublished paper, Yeshiva University, 1966. 21 pp. *D N T* (Copies from Dance Notation Bureau, 8 East 12th St., New York)

An example of the application of Rudolf Laban's effort-shape analysis to the study of the movement behavior in a family psychotherapy session. A sequence of movement and the accompanying speech is illustrated with words and effort notation. A recurrent pattern of movement "themes" observed in patient, therapist, and family members is described, with particular focus on how three distinct movement themes in the patient's movement parallel three phases in his verbal behavior: one coherent but depressed, one angry and grandiose, and one incoherent and disorganized.

194 **Davis, Martha.** "Movement Characteristics of Hospitalized Psychiatric Patients." In *Proceedings of the Fifth Annual Conference of the American Dance Therapy Association,* 1970. G T (Copies from ADTA, 5173 Phantom Court, Columbia, Md. 21043)

A Movement Diagnostic Scale developed from Laban's effort-shape analysis, containing over sixty-five items describing various types of "movement disturbance" seen in hospitalized schizophrenic patients, was used in a study of twenty-two patients of different diagnoses. Patients were observed behind a one-way screen in psychotherapy sessions with no sound. There was high observer reliability and a high correlation between presence of movement "fragmentation" and a history of more than two hospitalizations. There was also evidence of a relation between high phenothiazine medication and types of "reduced mobility." Patients who had similar movement factors and profiles derived from the diagnostic scale appeared, on study of their psychiatric records, to have similar diagnostic features. There was also preliminary evidence of the scale's potential for measuring improvement.

195 **Davis, Martha, and Schmais, Claire.** "An Analysis of the Style and Composition of 'Water Study'." In *Research in Dance: Problems and Possibilities,* pp. 105-113. New York: Committee on Research in Dance, 1967. D (Copies from CORD, New York University, 675 Education Building, Washington Square, New York)

The authors set out to analyze a film of a dance by Doris Humphrey in the way a music theorist might study a musical composition. Using terms developed by Laban for describing movement, they selected those movement variables which appeared most important in the dance, observed how they occurred in patterns, and discovered how the dance was composed of a hierarchy of increasingly larger patterns resembling "waves," intricately and harmoniously interrelated.

196 **Davis, R. C.** "The Specificity of Facial Expressions." *Journal of General Psychology* 10 (1934):42-58. T

Carney Landis' statistical analysis of subjects' facial expressions in different emotional situations is criticized, and a number of new statistical analyses are performed. They show that, contrary to Landis's conclusions, there are facial expressions specific to situations. The author assesses which muscles show the greatest degree of specificity.

197 **Davis, R. C.** "The Relation of Muscle Action Potentials to Difficulty and Frustration." *Journal of Experimental Psychology* 23 (1938):141-158. G T

This experiment shows that the more difficult the mental task, the greater the increase in muscular activity. Apparatus for measuring muscle potentials from arm and neck is described, along with the test procedures and results.

198 **Davis, R. C.** "Methods of Measuring Muscular Tension." *Psychological Bulletin* 39 (1942):329-346.

A valuable review of physiological and psychological literature on muscle tension, ways that it has been defined and a description and evaluation of various devices for measuring it.

199 **Davis, R. C., and Buchwald, Alexander M.** "An Exploration of Somatic

Response Patterns: Stimulus and Sex Differences." *Journal of Comparative and Physiological Psychology* 50 (1957):44-52. *G T*

Twenty-four men and women were shown a series of pictures while various measures were obtained (EMG, GSR, breathing pattern, etc.). The males were found to respond more "distally," the females more in "axial regions." The male subjects' responses differed according to the stimuli.

200　**Davitz, Joel R.** *The Communication of Emotional Meaning.* New York: McGraw-Hill Book Company, 1964. 214 pp. *G T*

This is a collection of studies of emotional communication, most of them dealing with vocal patterns. However, two chapters deal with facial expression: Chapter 2 by Davitz, "A Review of Research Concerned with Facial and Vocal Expressions of Emotion"; Chapter 7 by Eugene A. Levitt, "The Relationship between Abilities to Express Emotional Meanings Vocally and Facially."

201　**Dawson, William W., and Edwards, R. W.** "Motor Development of Retarded Children." *Perceptual and Motor Skills* 21 (1965):223-226. *G*

When subjects were matched for height and weight as well as sex, socioeconomic status, and other factors, there was no difference in grip strength, suggesting that physiological development accounts for differences found in the strength of normal and retarded children.

202　**Dayal, Leela Row.** *Manipuri Dances.* London: Oxford University Press, 1951. 53 pp. *D*

A little book describing and illustrating with drawings each movement of specific dances from Manipur, India, with notes on their meaning.

203　**De Jong, Russell N.** "Abnormal Movements." In *The Neurologic Examination*, pp. 499-519. New York: Paul B. Hoeber, 1950. *P* [Latest edition: 1967]

A very clear and comprehensive presentation of how motor disturbances are defined and diagnosed by the neurologist. Various types of tremors, spasms, tics, convulsions, fibrillations, myokymic, myoclonic, choreiform, athetoid, and dystonic movements are described and discussed with relation to body parts and muscles involved, the context in which they occur, pattern of occurrence, organic or psychogenic factors, diagnosis, and in some cases treatment. While patterns derived from electromyography are discussed, most of the analysis is based on direct observation of the body movement and uses descriptive and/or kinesiological terminology.

204　**De Kleen, Tyra.** *Mudrās: The Ritual Hand-poses of the Buddha Priests and the Shiva Priests of Bali.* New York: E. P. Dutton & Company, 1924. 62 pp. *D* [New Hyde Park, N.Y.: University Books, 1970]

Following a discussion of the history and background of the *mudrās*, there are a large number of drawings of the hand poses and a description of when they occur and what they symbolize.

205　**Dell, Cecily.** *A Primer for Movement Description: Using Effort-Shape and Supplementary Concepts.* New York: Dance Notation Bureau, 1970. 123 pp. *N*

An extensive presentation of the effort-shape system for analyzing and notating movement quality and style first developed by Rudolf Laban. This primer is written mainly for students of effort-shape analysis and those with experience in dance, body movement training, or systematic movement observation. It is replete, however, with examples from many contexts, which reflect the wide application of the system and which are readily comprehensible also to someone not trained in dance or body movement analysis. Concepts elucidated include the variables of "effort" and "shape," spatial orientation, body part involvement, and body attitude. Methods of notating patterns of movement are discussed.

206 **Delman, Louis**. "The Order of Participation of Limbs in Responses to Tactual Stimulation of the Newborn Infant." *Child Development* 6 (1935):98-109. *T*

A film analysis of "mass activity" in newborns showed some pattern regarding which are the first and second limbs to move.

207 **De Mille, Agnes**. *The Book of the Dance*. New York: Golden Press, 1963. 252 pp. *D P* [Racine, Wis.: Weston Publishing Company, 1968]

Brief historical and interpretive discussion of a wide range of dances, including the development of Western dance, a comparison of Eastern and Western dances, twentieth-century ballet, and modern dance and choreography; lavishly illustrated with 400 drawings and photographs.

208 **Denby, Edwin**. *Looking at the Dance*. (1949; reprint ed., New York: Horizon Press, 1968). 432 pp.

This large collection of articles by a noted dance critic is a source of notes on dance aesthetics and the observation and interpretation of body movement.

209 **Dennis, Wayne**. "An Experimental Test of Two Theories of Social Smiling in Infants." *Journal of Social Psychology* 6 (1935):214-223.

A pair of infant twins were raised for a period under controlled conditions. Two people cared for them but did not talk to, play with, or smile at the infants for seventy-six days. Nevertheless the infants developed smiling upon seeing the adult. The authors consider that the smile becomes "a conditioned response to any stimulus which brings about cessation of fretting, unrest, and crying" (p. 221).

210 **Dennis, Wayne, and Dennis, Marsena G**. "The Effect of Cradling Practices upon the Onset of Walking in Hopi Children." *Journal of Genetic Psychology* 56 (1940):77-86. *T*

Hopi Indian infants restricted to cradleboards and Hopi infants not put in cradleboards began walking at the same time, but both groups began two months later than other children studied.

211 **Dethier, V. G**. "Communication by Insects: Physiology of Dancing." *Science* 125 (1957):331-336. *D*

An analysis of the stereotyped "dances" of the fly, the conditions for their occurrence, and their similarities to "bee dances," which inform other bees where food has been found.

212 **Deutsch, Felix**. "Analysis of Postural Behavior." *Psychoanalytic Quarterly* 16 (1947):195-213.

Following a brief review of the literature and a psychoanalytic discussion of movement behavior, Deutsch presents nine cases with descriptions of the movements, positions, and accompanying verbal themes interpreted from a psychoanalytic point of view. To collect these rich, complex records, Deutsch made "posturograms" of the positions of the limbs and the body posture, along with notes on verbal themes, throughout six months to three years of analytic sessions.

213 **Deutsch, Felix**. "Thus Speaks the Body: An Analysis of Postural Behavior." *Transactions of the New York Academy of Sciences* 12 (1949):58-62.

A summary of Deutsch's theories about the function and meaning of various positions and movements he observed and recorded along with verbal comments during the analytic sessions of twenty-eight patients. He notes that individuals have consistent characteristic postures (here "posture" refers to each position of limbs, head, and trunk); that these positions change when there is psychodynamic change; that the body movement may anticipate impulses and thoughts which are not yet conscious; and that the positions can have highly specific symbolic meanings.

214   Deutsch, Felix. "Thus Speaks the Body: IV, Some Psychosomatic Aspects of the Respiratory Disorder: Asthma." *Acta Medica Orientalia* 10 (1951):67-86. *T*

A fascinating detailed analysis of breathing patterns, including reference to prenatal patterns, sex differences, normal breathing, individual differences, changes during emotional states (particularly anxiety), and the breathing patterns of asthmatics, with focus on one patient who was in analysis with the author. Using the Benedict-Roth respiratory apparatus and obtaining recordings with the spirograph and the pheumograph, Deutsch derived measures not only of the rhythm of respiration but also of the movement patterns of chest and abdomen as well. Arguing that breathing patterns reflect personality, Deutsch discusses his observations within a psychoanalytic framework.

215   Deutsch, Felix. "Analytic Posturology." *Psychoanalytic Quarterly* 21 (1952): 196-214.

As in earlier papers, Deutsch presents cases (five) of the motor behavior and psychodynamics observed during several years or months of analysis. This paper stresses theoretical formulations and specific interpretations of movements and positions: e.g., an upward, inward position of the feet reflects resistance against passive tendencies, or "any arrested movement is unconscious inhibition." In most of the interpretations, there is an intricate relationship between the motor behavior and the patient's early experiences, identifications, sexual fantasies, etc.

216   Deutsch, Felix. "Correlations of Verbal and Nonverbal Communication in Interviews Elicited by the Associative Anamnesis." *Psychosomatic Medicine* 21 (1959):123-130.

A woman with psychosomatic symptoms was interviewed by three different psychiatrists, and verbal recordings and observations of postural configurations and actions were made. The verbal content and body movement observations of the three interviews were contrasted, and specific words and accompanying movements were correlated. This clinical study illustrates interesting differences in the kind and degree of movement that occurred in the three sessions and the type of psychoanalytic interpretations Deutsch made of specific gestures and actions, right-left alterations, and nonverbal interactions. A summary of his theories concerning the nature and origin of these movements is given.

217   Deutsch, Felix. "Some Principles of Correlating Verbal and Non-verbal Communication." In *Methods of Research in Psychotherapy*, edited by L. A. Gottschalk and A. H. Auerbach, pp. 166-184. New York: Appleton-Century-Crofts, 1966. *T*

Deutsch summarizes his principles of interpretation of body movement and provides notes on the patient's history. There follows a fifteen-page transcript of a segment of a psychotherapy session in four columns: the verbal dialogue, notes on the patient's movements (e.g., "bends forward," "right hand outward"), the therapist's movements, and Deutsch's psychoanalytic interpretations of the verbal and nonverbal behavior.

218   Devereux, George. "Some Mohave Gestures." *American Anthropologist* 51 (1949):325-326.

A brief list of some stylized Mohave hand gestures that usually accompany speech, with notes on when they occur or what they refer to.

219 **Devereux, George**. "Mohave Indian Verbal and Motor Profanity." *Psychoanalysis and the Social Sciences* 3 (1951):99-127.

Includes description and psychoanalytic interpretation of a number of obscene gestures and sexual behaviors among the Mohave Indians.

220 **De Vore, Irven**, ed. *Primate Behavior: Field Studies of Monkeys and Apes.* New York: Holt, Rinehart and Winston, 1965. 654 pp. *D G P T*

One of the important collections of primate field studies, this work has a great deal of information about facial and body "displays," patterns of group interaction, mother-infant interaction and developmental stages, sexual activity and courting behaviors, locomotion patterns, and dominance-submission behaviors. Of particular note for the detail and richness of the body movement descriptions are Chapter 7 by Phyllis Jay, on the common langur; Chapter 11 by Vernon and Frances Reynolds, on Budango Forrest chimpanzees; and Chapter 12 by Jane Goodall, on the Bombe Stream chimpanzees. Also noteworthy are Chapter 3 by Hall and De Vore, Chapter 15 by Mason, and Chapter 16 by Marler, which are separately reviewed here.

221 **Dewey, Evelyn.** *Behavior Development in Infants: A Survey of the Literature on Prenatal and Postnatal Activity, 1920-1934.* New York: Columbia University Press, 1935. 321 pp.

An excellent review of research done by Gesell, Buhler, McGraw, and others on infant behavior and motor patterns. Studies of the development of reflexes, locomotion, prehension, visual behavior, etc., up to two years of age are surveyed, and sections on animal and human fetal activity are included. Because this book covers the major period of such research, it is still very timely and valuable.

222 **Dickey, Elizabeth C., and Knower, Franklin H.** "A Note on Some Ethnological Differences in Recognition of Simulated Expressions of the Emotions." *American Journal of Sociology* 47 (1941):190-193. *T*

A report of a study indicating that Mexican schoolchildren judged facial expressions of emotion more accurately than American schoolchildren did.

223 **Diehl, Katherine S.** *Religions, Mythologies, Folklores: An Annotated Bibliography.* 2d ed. Metuchen, N.J.: Scarecrow Press, 1962. 573 pp.

A 2,388-title bibliography of religious and folklore literature published between 1900 and 1960, with brief annotations and subject indexing. It is relevant here as a source of references on rituals, ceremonies, and manners of such diverse cultures and groups as Pueblo Indians, Gypsies, and Indian Hindus.

224 **Dierssen, Guillermo; Lorenc, Mary; and Spitaleri, Rose M.** "A New Method for Graphic Study of Human Movements." *Neurology* 11 (1961):610-618. *D G P*

Describes a technique for accurate measurement of body movements from frames of films: two points on the chair arms are used as fixed reference points, and vertical and lateral coordinates of a point or points on the moving limb are recorded relative to these references. Developed for measuring complex finger or arm movements in neurological disorders, the technique results in a graph of the movement's spatial path over time.

225 **Dincmen, Kriton.** "Chronic Psychotic Choreo-Athetosis: A Clinical Study of the Subject." *Diseases of the Nervous System* 27 (1966):399-402. *T*

The author posits a new disease entity based on a survey of movement patterns of 1,700 long-term psychiatric patients. Although there is predominantly psychotic symptomatology, it is considered a neurological disease with striatal symptoms such as athetosis and involuntary movements. It is unclear, however, how patients with "chronic psychotic choreo-athetosis" differ from chronic schizophrenic patients with abnormal movement.

226   **Ding, Gladys F., and Jersild, Arthur T**. "A Study of the Laughing and Smiling of Preschool Children." *Journal of Genetic Psychology* 40 (1932):452-472. *T*

The incidence of laughter and smiling, together with the situations that evoked it, of fifty-nine Chinese children in an American nursery school was recorded over 276 hours of observation. Marked individual differences but no correlations with age, physical build, or socioeconomic status were noted.

227   **Dingman, Harvey F., and Silverstein, Arthur B**. "Intelligence, Motor Disabilities, and Reaction Time in the Mentally Retarded." *Perceptual and Motor Skills* 19 (1964):791-794. *T*

Contrary to prior research showing that retarded individuals have longer reaction times than normals, the authors conducted a study involving tests of steadiness, tapping, and reaction time that showed no correlation between intelligence and reaction time in these patients when the effects of motor disability were controlled for.

228   **Dittmann, Allen T**. "The Relationship between Body Movements and Moods in Interviews." *Journal of Consulting Psychology* 26 (1962):480.

Correlation was found between the incidence of movements of specific body parts and the patient's professed mood, as studied from filmed interviews. When he was angry, for example, there were many head and leg, but few hand, movements.

229   **Dittmann, Allen T**. "Kinesic Research and Therapeutic Processes: Further Discussion." In *Expression of the Emotions in Man*, edited by P. H. Knapp, pp. 140-147. New York: International Universities Press, 1963.

Extensive study of filmed psychoanalytic sessions failed to show a relationship between speech disturbances and foot movements or any general expressive properties of foot movements.

230   **Dittmann, Allen T**. "The Body Movement-Speech Rhythm Relationship as a Cue to Speech Encoding." In *Studies in Dyadic Communication*, edited by A. W. Siegman and B. Pope. New York: Pergamon Press, forthcoming.

A review of Dittmann and Boomer's research on speech and movement in interviews, particularly film study of the relation between accent and amount of movement and phonemic clauses and hesitations.

231   **Dittmann, Allen T., and Llewellyn, Lynn G**. "Relationship between Vocalizations and Head Nods as Listener Responses." *Journal of Personality and Social Psychology* 9 (1968):79-84. *T*

Pairs of subjects spoke to each other while wearing microphones and transducers for measuring head nods (this equipment is described). The head nods were found to occur significantly with vocalizations such as "um hum" or "I see," following a talker's speech. Analysis of the data suggests that these behaviors indicate that the listener wishes to speak or the talker needs feedback. Typically, a listener may nod, then vocalize when the speaker has finished.

232   **Dittmann, Allen T.; Parloff, Morris B.; and Boomer, Donald S**. "Facial and Bodily Expression: A Study of Receptivity of Emotional Cues." *Psychiatry* 28 (1965):239-244. *T*

Using film segments of a woman being interviewed in which the authors decided the face and body both expressed pleasant or unpleasant affect or differed, psychotherapists and dancers were asked to rate the segments for pleasant or unpleasant affect from seeing either the whole body or the body without the face. There were significant differences between therapists and dancers; the dancers tended to score more unpleasant affect and responded more to bodily cues. A noted dance therapist judged the films exactly opposite to the therapists, apparently responding primarily to bodily cues.

233 **Dorcy, Jean.** *The Mime.* Translated by R. Speller, Jr., and P. de Fontnouvelle. New York: Robert Speller and Sons, 1961. 116 pp. *P*

A beautifully illustrated book on the nature of modern mime—and within it the nature of man and human movement—with brief articles by such great mime artists as Barra it, Decroux, and Marceau.

234 **Down(, June E.** *The Will-Temperament and Its Testing.* Yonkers, N.Y.: World Boo Company, 1923. 339 pp. *G T*

"Will-te nperament" and personality characteristics are determined here from tests of t1e individual's pattern of motor impulsion or inhibition, as observed from his handwriting and the way he moves while writing. Profiles based on scores for speed of movement, flexibility, motor impulsion, coordination of impulses, etc., are presented and interpreted. The psychological significance of each factor is discussed. Profiles of psychotic subjects are analyzed, and the relationship between temperament patterns and intelligence is explored.

235 **Drag, Richard M., and Shaw, Marvin E.** "Factors Influencing the Communication of Emotional Intent by Facial Expressions." *Psychonomic Science* 8 (1967):137-138. *T*

This study replicated Thompson and Meltzer's findings that subjects can best communicate "happiness" and have the most difficulty with "contempt." Additional findings were that women communicated certain affects more effectively than men and that measures of anxiety did not correlate with enactment ability.

236 **Dratman, Mitchell L.** "Reorganization of Psychic Structures in Autism: A Study Using Body Movement Therapy." In *Proceedings of the Second Annual Conference of the American Dance Therapy Association,* pp. 39-45, 1967. (Copies from ADTA, 5173 Phantom Court, Columbia, Md. 21043)

A description of how the dance therapist works with the autistic child and a discussion of the function of body movement therapy in developing body awareness, the beginnings of relatedness, and a sense of self in these children.

237 **Drillis, Rudolf.** "Objective Recording and Biomechanics of Gait." *Annals of the New York Academy of Sciences* 74 (1958):86-109. *D G P*

Concerned with the objective measurement of gait in physical medicine, the author reviews and illustrates a number of instruments and techniques and summarizes data on kinesiological and mechanical aspects of normal and pathological gait. His bibliography is useful for sources on the kinesiology of gait. In this paper the following are discussed: (1) a technique for measuring walking time and stride with electrical equipment and oscillograph; (2) naturalistic observation of the walking speed of 936 normal people; (3) kinematic techniques for recording and analyzing gait path, stride, speed, etc.; (4) three-dimensional photography of gait: (5) techniques for analyzing the path of the center of gravity and the force of the step.

238   **Druckman, Ralph; Seelinger, Donald; and Thulin, Barbara**. "Chronic Involuntary Movements Induced by Phenothiazines." *Journal of Nervous and Mental Disease* 135 (1962):69-76. *P*

Case histories of psychiatric patients who developed severe involuntary movements of mouth, limbs, or trunk following phenothiazine medication are described. Older patients and those who have had EST may be predisposed to "dyskinesia," which may not stop even after medication is withdrawn.

239   **Dudek, S. Z.; Lester, L. P.; and Harris, B. R**. "Variability on Tests of Cognitive and Perceptual-Motor Development in Kindergarten Children." *Journal of Clinical Psychology* 23 (1967):461-464. *T*

Comparison of results for 107 five- and six-year-olds on the WISC, the Lincoln Oseretsky Motor Development Scale, and the Rutgers Drawing Test with a psychiatric assessment of their normalcy and home environment showed that there was little variability on the motor and performance tests but a great deal on the verbal I.Q. test, which correlated with "maturational hazards in the home." The motor tests correlated with each other and significantly with I.Q.

240   **Duffy, Elizabeth**. "Tensions and Emotional Factors in Reaction." *Genetic Psychology Monographs* 7 (1930):1-79. *G T*

The degree of hand-muscle tension was measured in children from three and one-half to five years old through pressure on a rubber bulb held in one hand during performance of key presses to lights by the other hand. Kymographic records were obtained. Ratings of the child's degree of "excitability" correlated significantly with his height-of-tension scores. Tension lines were highly individual, and certain features such as high irregular lines correlated dramatically with clinical information about the child's personality (e.g., impulsive and distractible).

241   **Duffy, Elizabeth**. "Muscular Tension as Related to Physique and Behavior." *Child Development* 3 (1932):200-206.

Continuation of a study reported earlier on muscular tension patterns in nursery children, measured with dynamographs. Slight relationships were found between high muscular tension and low body weight, slight body build, higher pulse rate, poorer motor performance, less play activity, avoidance of physical contact, use of few words, and more distractibility.

242   **Duffy, Elizabeth**. "Level of Muscular Tension as an Aspect of Personality." *Journal of General Psychology* 35 (1946):161-171. *T*

The muscle tension characteristics of twenty-five women were tested in three ways while they performed certain mental and motor tasks on different occasions. A factor analysis yielded factors of general levels of tension and point pressure that were individually consistent over time, as well as two other factors specific to the task or mode of measurement.

243   **Duffy, Elizabeth**. *Activation and Behavior*. London: John Wiley & Sons, 1962. 384 pp.

Activation, or "the range of physiological changes from deep sleep to extreme excitement" (p. 4), is explored in relation to its measurement, patterning, and underlying influences; sensory and motor correlates; and individual differences in activation. A great deal of literature on muscle tension, motor performance, and activation level is discussed, including muscle potential, reaction time, and speed and intensity of response studies. Abnormalities of breathing and muscle tension in schizophrenic and neurotic patients are discussed. The book includes an excellent forty-page bibliography.

244 **Dunbar, Flanders**. *Emotions and Bodily Changes: A Survey of Literature on Psychosomatic Interrelationships 1910-1953*. 4th rev. ed. New York: Columbia University Press, 1954. 1192 pp.

This monumental work includes a chapter on interrelationships of emotion, muscle tension, and body movement with somatic illness such as arthritis.

245 **Dunbar, Flanders**. "Interpretation of Body Behavior during Psychotherapy." In *Science and Psychoanalysis,* vol. 3, *Psychoanalysis and Human Values,* edited by J. H. Masserman, pp. 223-230. New York: Grune & Stratton, 1960.

A discussion of the significance of body movement behavior in psychoanalysis, with particular focus on characteristics of "muscle-bound" patients and the necessity of their developing greater relaxation and body awareness if deeper therapy is to be effective.

246 **Duncan, Isadora**. *The Art of the Dance*. 2d ed. Edited by S. Cheney. New York: Theatre Arts Books, 1928. 147 pp. *D P*

A beautifully illustrated book containing eulogies to the artist and articles by her on the aesthetics of dance in relation to nature, emotion, and classical history.

247 **Duncan, Starkey, Jr**. "Nonverbal Communication." *Psychological Bulletin* 72 (1969):118-137.

A concise review of recent research in paralanguage, kinesics or body motion, and proxemics, divided into those who use a "structural" or systems approach analogous to language analysis (e.g., Birdwhistell, Scheflen, Condon and Ogston, Hall) and those who statistically correlate nonverbal variables with "external" variables (e.g., Ekman and Friesen, Sainsbury, Dittman, Parloff and Boomer, Charny, Exline, Argyle and Dean, Kleck, Kendon).

248 **Duncan, Starkey, Jr**. "Towards a Grammar for Floor Apportionment: A System Approach to Face-to-Face Interaction." In *Proceedings of the 2nd Annual Environmental Design Research Association Conference*, pp. 225-235. Philadelphia: Environmental Design Research Association, 1970. *T* (Copies from author, c/o Department of Psychology, University of Chicago, 5848 South University Avenue, Chicago, Ill. 60637)

Extensive recording and analysis of the language, paralanguage, and body motion behavior of therapist and client in an initial psychiatric interview was done to determine the rules which regulate the flow of interaction and the "smooth exchange of speaker and auditor roles." Behavioral cues of "floor-yielding" and "floor-retaining" and the rules for their occurrence are described. In the case of body motion a ceasing or relaxing of the gesture is a cue to floor-yielding.

249 **Duncan, Starkey, Jr**. "Towards a Grammar for Taking Speaking Turns in a Conversation: A System Approach to Face-to-Face Interaction." Unpublished paper, University of Chicago, [1970]. 73 pp. *D G T* (Copies from author, c/o Department of Psychology, University of Chicago, 5848 South University Avenue, Chicago, Ill. 60637)

An extensive report on the author's videotape analysis of "turn-taking" between an interviewer and interviewee. Detailed descriptions of paralinguistic, visual, and body motion signals for turn-taking and the system of rules governing it, together with the notations used and quantitative data, are presented.

250 **Dunlap, Knight**. "The Role of Eye-Muscles and Mouth-Muscles in the Expression of the Emotions." *Genetic Psychology Monographs* 2 (1927):195-233. *P*

Judges were shown blocks of photographs, two originals of people responding to various stimuli and two composites made up of the lower part of one original and the upper part of the other. First they decided what emotions were expressed in the "natural" photographs, then in the composites. Results were interpreted as evidence that the mouth area is more important than the eyes in the expression of pleasure or displeasure.

251   Dyer-Bennett, Melvene. "Some Thoughts about Change—Our Most Deceptive Therapeutic Goal." In *Proceedings of the Fifth Annual Conference of the American Dance Therapy Association* 1970. (Copies from ADTA, 5173 Phantom Court, Columbia, Md. 21043)

An eminent motility therapist discusses the nature of change in psychotherapy, movement as a medium for change, and the "connection between awareness and action." She makes a distinction between "mobility" and "motility" and the potential of movement as either a direct expression of feeling or a defense against it.

252   Edwards, Nancy. "The Relationship between Physiological Condition Immediately after Birth and Mental and Motor Performance at Age Four." Ph.D. dissertation, Columbia University, 1966. 101 pp. (Datrix order no. 67-6523)

The physical status of 147 neonates as measured by birth weight and Apgar scores was compared with their performance on tests of intelligence, conceptual ability, and fine and gross motor coordination at age four. The correlations between Apgar scores at birth and the mental and motor tests at four years were found to be highly consistent and positive. The highest correlations were between Apgar scores and motor coordination, and further statistical analysis revealed that the Apgar rating of muscle tone was the single best predictor.

253   Efran, Jay S. "Looking for Approval: Effect on Visual Behavior of Approbation from Persons Differing in Importance." *Journal of Personality and Social Psychology* 10 (1968):21-25. *T*

Male freshman subjects and two confederates (one said to be a freshman, one a senior) were asked to talk to each other for five minutes, then to fill out questionnaires about their impressions. The subject's gaze behavior was recorded by an observer with an Esterline-Angus multipen recorder. In each case one confederate was instructed to smile and look more (i.e., "approve"), the other to be attentive but neutral. Results show that (1) the subjects felt better liked by the "approving" person and (2) they looked more at the approving confederate, particularly if he was of higher status.

254   Efron, David. *Gesture and Environment.* New York: King's Crown Press, 1941. 184 pp. *D N T* [to be published in a new edition entitled *Gesture and Culture,* with an introduction by Paul Ekman. The Hague: Mouton Publishers]

The subtitle of this work is "A tentative study of some of the spatio-temporal and 'linguistic' aspects of the gestural behavior of Eastern Jews and Southern Italians in New York City, living under similar as well as different environmental conditions." Following a fascinating review of historical literature and German theories on the biological basis of gesture, body type, and cultural temperament, Efron offers an empirical refutation of the racist theories. Through observation, sketches, and motion pictures of "traditional" and "assimilated" Jews and Italians in their own settings, Efron shows that with cultural assimilation body movements change, and those of Jews and Italians begin to resemble each other in "hybrid gestures." He also notes changes as they move upward socio-economically. The way he analyzes and graphically records movements is innovative. He studied the "radius," number, body parts used, form or path, tempo, quality, patterns of laterality, and planes of the gestures; the relationship between speech and "ideographic" versus "physiographic" gestures; how close people stand, how objects are used, and what one does in gesture. How each of the four groups differ in movement characteristics is described. The wealth of information presented in text, drawings, and notations has yet to be developed with the vision and breadth that Efron displays here.

255 **Ehrlich, Milton P.** "The Role of Body Experience in Therapy." *Psychoanalytic Review* 57 (1970):181-195.

Following a sketch of attitudes toward the significance of body movement and awareness as stated by analysts and movement therapists, the author presents three brief cases with patients' reports of their experience of their bodies and physical movement in therapy.

256 **Eibl-Eibesfeldt, Irenaus.** *Ethology: The Biology of Behavior.* New York: Holt, Rinehart and Winston, 1970. 530 pp. *D G P*

Comprehensive and beautifully presented, this book outlines and reviews major research in ethology and the comparative behavior of animals and man in the tradition of Lorenz. As such it deals throughout with analysis of actions, movements, and expressive and communicative behaviors from insects to the primates. Of particular note are the chapters on "fixed action patterns," the origins and functions of expressive movements and signals, behavior towards species members, temporal and spatial factors of animal behavior, and human social behavior, particularly cross-cultural study of greetings. It concludes with an extensive bibliography of ethology literature, much of it in German.

257 **Eibl-Eibesfeldt, Irenaus.** "Transcultural Patterns of Ritualized Contact Behavior." In *Behavior and Environment: The Use of Space by Animals and Men,* edited by A. H. Esser, pp. 238-246. New York: Plenum Publishing Corporation, 1971. *P*

The author presents observations of greeting behavior, shown to be the same cross-culturally from Europe to Brazil to Samoa, and discusses the innate versus the learned nature of such patterns. The greeting pattern consists of a smile and a rapid raising of the eyebrows. The relation of this pattern to similar animal behaviors and its functional significance are discussed.

258 **Eisenberg, Philip.** "Expressive Movements Related to Feeling of Dominance." *Archives of Psychology* 211 (1937):1-73. *T*

On the basis of self-ratings and the Social Personality Inventory, two groups of men and women were selected: thirty-three with high dominance feelings (self-confidence, superiority, etc.) and twenty-nine who indicated feelings of nondominance, shyness, self-consciousness, etc. The subjects chosen were observed in a number of structured tasks such as reading aloud, writing, and eye fixations. A great many quantitative and qualitative differences in the behavior of the two groups are reported, from manner of sitting and walking to facial expressions. Two women, one dominant and one nondominant, are described in detail to illustrate the dramatic differences. Certain differences between the sexes are also noted.

259 **Eisenberg, Philip, and Reichline, Philip B.** "Judging Expressive Movement: II, Judgement of Dominance-Feeling from Motion Pictures of Gait." *Journal of Social Psychology* 10 (1939):345-357. *T*

Based on prior tests of dominance feeling, eight "extremely dominant" and eight "extremely nondominant" women were filmed while walking. Judges were asked to distinguish the dominant from the nondominant. Their judgments were a little better than chance; there were no sex differences in observer judgment, and the better judges tended to focus more on subtler movement details. The ability and influences of the observers are extensively analyzed.

260 **Ekman, Paul.** "A Methodological Discussion of Nonverbal Behavior." *Journal of Psychology* 43 (1957):141-149.

Since Ekman is now one of the most prominent and productive researchers of nonverbal behavior, this paper is of historical interest. In it he clearly sets out to use the experimental method, sampling techniques, and frequency measures of nonverbal categories to be selected by the investigator. He discusses recording methods such as a twenty-channeled operations recorder or sampling still photos from motion pictures.

261 **Ekman, Paul.** "Body Position, Facial Expression and Verbal Behavior during Interviews." *Journal of Abnormal and Social Psychology* 68 (1964):295-301. *T*

To assess the communicative value of body position and facial expression, photographs were taken every fifteen or thirty seconds and tape recordings were made from two standardized stress interviews. Written speech samples together with pairs of photos, one of which was actually from the given speech, were presented to different groups of judges (young students, dancers, a class of students from ages eighteen to fifty-five). The judges picked the correct photo to a significant degree when the whole body or just the head was shown. However, when the body alone was shown without the head, their judgments were accurate for only one interview. Ekman discusses the possibility that individuals differ in "sending behavior," that is, how and when they communicate certain things with their bodies.

262 **Ekman, Paul.** "Differential Communication of Affect by Head and Body Cues." *Journal of Personality and Social Psychology* 2 (1965):726-735. *T*

Four experiments in which subjects rated photographs either of face, of body only, or of face and body on three dimensions of affect as defined by Schlosberg: pleasantness-unpleasantness, attention-rejection, and sleep-tension. Results showed that the head and face gave information indicating what affect was expressed—in contrast to the body, which communicated the degree of intensity of the affect.

263 **Ekman, Paul.** "Communication through Nonverbal Behavior: A Source of Information about an Interpersonal Relationship." In *Affect, Cognition and Personality*, edited by Silvan S. Tomkins and Carroll E. Izard, pp. 390-442. New York: Springer Publishing Co., 1965. *P T*

After a brief review of psychology literature on nonverbal behavior, eleven controlled experiments are reported, using photos taken from interviews with both a stress and a relief phase. Ekman investigates whether observers agree in their interpretations of the photos under various conditions—e.g., given only one photo, given photos with both interviewer and interviewee or only with the interviewee, etc. The results generally support the hypothesis that individuals can get accurate information from nonverbal behavior as perceived in photos. Consistency and accuracy of judgment are studied further. Then, in a third series of experiments, subjects rated the photos on three affective dimensions: pleasantness-unpleasantness, attention-rejection, and sleep-tension. Results showed that ratings of "unpleasant" were consistently given for the stress phase and "pleasant" for the cathartic phase. Ekman concludes with a discussion of possible ways to measure individual differences and comments on the results and significance of the experiments.

264 **Ekman, Paul**. "Universals and Cultural Differences in Facial Expressions of Emotion." In *Nebraska Symposium on Motivation*. Lincoln: University of Nebraska Press, forthcoming.

265 **Ekman, Paul**, ed. *Darwin and Facial Expression*. New York: Academic Press, forthcoming.

266 **Ekman, Paul, and Friesen, Wallace V.** "Head and Body Cues in the Judgment of Emotion: A Reformulation." *Perceptual and Motor Skills* 24 (1967): 711-724. *T*

Using photographs from five stress interviews showing either head alone or the body without the head, judges were to rate them according to Woodworth's six affects. There was greater agreement on the pictures that showed acts rather than positions and on those with head alone, especially for judgments of happiness, surprise, fear, and contempt. The authors discuss the notion that, when emotionally aroused, a person will move rather than stay still and that different types of cues (i.e., acts, positions, facial expressions, and head orientation) give different information concerning what the affect is or how intense it is.

267 **Ekman, Paul, and Friesen, Wallace V.** "Nonverbal Behavior in Psychotherapy Research." In *Research in Psychotherapy*, vol. III, edited by J. M. Shlien, pp. 179-216. Washington, D.C.: American Psychological Association, 1968. *P T*

Throughout this chapter Ekman and Friesen discuss a number of theoretical and methodological issues, such as why one should study nonverbal behavior, , what kinds of information can be obtained, and how a unit of behavior should be defined. They report on research with silent films of depressed women patients at the time of admission and at discharge, showing agreement between judges as to affect expressed, interpersonal style, and individual differences, among others. Departing from earlier work using still photographs, Ekman and Friesen define the unit of study as roughly the beginning to end of a movement and discuss "indicative" (direct) and "communicative" (based on observer judgment) methods of determining an act's meaning. They report research on the film of one patient that shows change in frequency and agreement in interpretation of foot and hand movements, as well as consistency in the verbal content accompanying a specific gesture.

268 **Ekman, Paul, and Friesen, Wallace V.** "A Tool for the Analysis of Motion Picture Film or Video Tape." *American Psychologist* 24 (1969):240-243.

The authors describe the VID-R (Visual Information Display and Retrieval) system for collecting, ordering, and abstracting information from videotapes or films converted into tapes. The system utilizes a set of recorders, monitors, printers, and a digital computer. An operator can locate and collect bits of behavior to be shown at various speeds by specifying time, location, or content (e.g., "all hand movements.")

269 **Ekman, Paul, and Friesen, Wallace V.** "Nonverbal Leakage and Clues to Deception." *Psychiatry* 32 (1969):88-105. *T*

Following a long discussion of aspects of the deceptive situation where information is withheld or lied about to oneself or another, the authors hypothesize about the different "sending capacities" or the body, that the face will "lie" the most, the hands next, and the feet least, because they differ anatomically in terms of the degree of mobility and socially in terms of how exposed or attended to they are. To test this, they used three filmed interviews of depressed women in which the researchers were sure deception or concealment of information had occurred. They showed them to two groups—one seeing only the face and head of the woman, and the other seeing only the body below the neck—who then filled out Adjective Check Lists as to the affect, personality features, etc., expressed. The results show that information about feelings which the patient withheld verbally was inferred from observing the body alone more than from watching the face.

270 Ekman, Paul, and Friesen, Wallace V. "The Repertoire of Nonverbal Behavior: Categories, Origins, Usage and Coding." *Semiotica* 1 (1969):49-98. *D*

An impressive effort at defining categories of nonverbal behavior and analyzing problems of perspective, interpretation, context, and origin. Following a succinct review of their research, Ekman and Friesen propose five categories of nonverbal behavior and discuss each with respect to its origins, its usage (the circumstances in which a nonverbal act occurs, including the person's intentions, the feedback he receives, and what type of information is conveyed: informative, interactive, or communicative), and its coding (the correspondence between the act and its meaning as either arbitrary, iconic, or intrinsic). The five categories are: emblems (actions with "agreed upon" meaning or a "dictionary definition"), illustrators (six types of body movement illustrating what is being said), affect displays (primarily facial expression), regulators (conversation pacers), and adaptors (actions derived from early adaptive behaviors related to needs, emotions, or interpersonal contacts). Research in progress on these categories is reported: cross-cultural study of emblems; examination of illustrators from films of psychiatric patients; research into pancultural interpretation of facial expression, using the Facial Affect Scoring Technique and study of the facial expression of individuals viewing stressful films; research on adaptors (i.e., how "self-adaptors" change as a psychiatric patient improves), the frequency of adaptors and linguistic phenomena, and cross-cultural differences in self-adaptors under stress.

271 Ekman, Paul, and Friesen, Wallace V. "Constants Across Cultures in the Face and Emotion." *Journal of Personality and Social Psychology* 17 (1971): 124-129. *T*

Members of a preliterate culture (the New Guinea Fore Group), virtually unexposed to literate cultures, were asked to choose which of two or three photographs of a face showing one of six expressions would go with a given story. Positive results support the hypothesis that certain facial expressions are universally associated with certain emotions.

272 Ekman, Paul; Friesen, Wallace V.; and Ellsworth, P. *Emotion in the Human Face.* New York: Pergamon Press, forthcoming.

273 Ekman, Paul; Friesen, Wallace V.; and Tomkins, Silvan S. "Facial Affect Scoring Technique: A First Validity Study." *Semiotica* 3 (1971):37-58. *P T*

Through a great deal of preliminary testing, a scoring technique that successfully predicts observer's judgments of facial expression was developed. The tool involves judgment of photographs or motion pictures, together with word descriptions of three areas of the face. A number of different-stimulus persons in posed expressions of six "single" emotions were used. The validity and application of the technique are reported, and observations of individual differences in facial expression are noted.

274   **Ekman, Paul; Sorenson, E. Richard; and Friesen, Wallace, V**. "Pan-Cultural Elements in Facial Displays of Emotion." *Science* 164 (Apr., 1969):86-88. *T*

To test the hypothesis that people from various cultures perceive the same "primary emotion" in a specific facial expression (although they may differ in what evokes the emotion, how its expression is modified, and what its consequences are), the authors presented photos of people showing the "pure display of a single affect": happiness, surprise, fear, anger, disgust-contempt, or sadness. Subjects from the United States, Brazil, Japan, New Guinea, and Borneo were to select the affect word from a list of six which they felt corresponded with the emotion expressed in the picture. Results corroborated the authors' predictions, although agreement was better in the literate groups.

275   **Ellis, Havelock.** "The Art of Dancing." In *The Dance of Life,* pp. 34-63. New York: Modern Library, 1929.

The function of dance in society is explored in a chapter that, despite its brevity, contains a great deal of information about dance cross-culturally and historically. The author focuses on the role of dance in religion, group organization, courtship, and socialization.

276   **Ellsworth, Phoebe C., and Carlsmith, J. Merrill**. "Effects of Eye Contact and Verbal Content on Affective Response to a Dyadic Interaction." *Journal of Personality and Social Psychology* 10 (1968):15-20. *T*

Following an interview in which the interviewer either looked very little at the subject (four times) or looked a great deal (twenty times) while discussing questions favorable or unfavorable to the subject, the subject filled out a questionnaire on the interviewer and the interview. Results showed that, if the topic was neutral or positive and the interviewer looked a lot, the subject liked her and the interview more; but if the topic was indirectly critical of the subject and the interviewer looked a lot, she liked the interviewer less than if she did not look a great deal. In the "Favorable/Look" and "Unfavorable/No Look" conditions the interviewer was also judged more "personal."

277   **Elworthy, Frederick T**. *The Evil Eye: An Account of This Ancient and Widespread Superstition.* London: John Murray (Publishers), 1895. 471 pp. *D* [New York: Collier Books, The Macmillan Company, 1970]

A history of superstitions and the belief in the power of certain actions, objects, and creatures to bring on or dispel evil. It has interesting passages on the superstitious use of touch and specific hand gestures in Europe and the symbolism of hand positions recorded in ancient art and literature. There are also illustrations of common gestures used in Italy in the nineteenth century.

278   **Engen, Trygg, and Levy, Nissim**. "Constant-sum Judgement of Facial Expression." *Journal of Experimental Psychology* 51 (1956):396-398. *T*

In an experiment where a subject's ratings of the expression shown in pictures were compared with Schlosberg's ratings on dimensions of expression, data was obtained showing that the constant-sum method used in psychometric scaling, which yields statistically more valuable ratio scales, could be used beyond psychophysical experiments for something like scaling facial expressions.

279   **Engen, Trygg; Levy, Nissim; and Schlosberg, Harold**. "The Dimensional Analysis of a New Series of Facial Expressions." *Journal of Experimental Psychology* 55 (1958):454-458. *T*

In three different experiments, subjects rated a new series of up to forty-eight pictures of facial expressions on a nine-point scale for each of the Schlosberg dimensions: pleasantness-unpleasantness, attention-rejection, and sleep-tension. High consistency and reliability of judgment are reported, with no clear transfer or practice effects between dimensions.

280   **Ernhart, Claire B.; Graham, Frances K.; Eichman, Peter L.; Marshall, Joan M.; and Thurston, Don**. "Brain Injury in the Preschool Child: Some Developmental Considerations: II, Comparison of Brain Injured and Normal Children." *Psychological Monographs*, Whole No. 574, 77 (1963):17-33. *G T*

On a battery of tests (among them, motor-coordination and perceptual-motor), seventy children with brain injury of differing etiology showed marked impairment in cognitive and perceptual-motor areas but not in personality. They also did not show a "hyperkinetic" syndrome.

281   **Escalona, Sibylle, and Heider, Grace Moore**. *Prediction and Outcome: A Study in Child Development*. New York: Basic Books, 1959. 318 pp. *G T*

Predictions about the infant's personality and cognitive and social development were made from extensive tests and observations of thirty-one infants and were compared with follow-up evaluations four to six years later. Among the areas evaluated were activity level, expressive behavior, and motor coordination and development; these tended to be consistent and accurately predicted. A great deal of the "working material" of this important study is included: data, scales, case description, and criteria and procedures for making predictions, assessing the children, and evaluating the results.

282   **Eshkol, Noa**. *The Hand Book: The Detailed Notation of Hand and Finger Movements and Forms*. Tel Aviv, Israel: Movement Notation Society, 1971. 135 pp. *D N* (Copies from Movement Notation Society, 75 Arlozorov Street, Holon, Israel)

Following an introduction to the Eshkol-Wachmann system of movement notation, its use in recording the hand movements of the sign language of the deaf and of Indian dancers is illustrated. Various positions of the hand are sketched next to the notation, and a glossary of words is presented in notation.

283   **Eshkol, Noa, and Wachmann, Abraham**. *Movement Notation*. London: Weidenfeld & Nicholson, 1958. 203 pp. *D N*

A system for recording movement which originated from dance but which is applicable to all forms of body movement. It involves a staff which vertically has twenty spaces for movements of right and left limbs, head and torso, weight placement, and front and which horizontally represents changes over time. There are symbols for rotary, plane, curved, or conical movements, and numbers signify the degree of the movement. Techniques for recording complex paths and special signs for symmetrical, passive, and so on, are other elements of the system.

284   **Espenak, Liljan**. "Body Dynamics and Dance as Supportive Techniques in Individual Psychotherapy." Unpublished paper, New York. 21 pp. (Copies from author, 2121 Broadway, New York 10023)

Following a discussion of what dance therapy is, case descriptions are presented of dance therapy with three neurotic patients focusing on their characteristic postures and areas of constriction, interpretation of intrapsychic conflicts and affects, and how the dance therapist worked with the patient through body movement.

285 **Espenak, Liljan**. "A New Non-verbal Approach to Personality Evaluation." Presented at American Association on Mental Deficiency Region 10 Conference, n.p., September 1970. 17 pp. *D* (Copies from author, 2121 Broadway, New York 10023)

This paper includes a drawing of the body with psychological interpretations of what specific areas relate to and a description of the author's Movement Diagnostic Tests in terms of aspects of personality and affect assessed, description of eight motor tasks given, and how these were scored and interpreted. A reprint of the rating form is included.

286 **Espenak, Liljan**. "Changing the Lifestyle through Psychomotor Therapy: A Case Presentation." Presented at the 11th Congress of the International Association of Individual Psychology, New York, July 1970. 11 pp. (Copies from author, 2121 Broadway, New York 10023)

Movement assessment and therapy with a young depressed woman is described.

287 **Estes, Stanley G**. "Judging Personality from Expressive Behavior." *Journal of Abnormal and Social Psychology* 33 (1938):217-236. *T*

An elaborate study in which some 323 judges evaluated the personalities of fifteen male subjects filmed in short structured situations, thereby basing judgment on manner and nonverbal behavior. Correlation of the judgments (which were made in the form of ratings, checklists, and matching procedures) with extensive clinical and experimental data on the filmed subjects' "true" personalities was done. "Success" depended on the judge, the subject, and the personality features being judged. There were marked differences in accuracy between judges who showed strong interest in painting and dramatics and judges who were faculty members. The analytic faculty members did poorly. The filmed subjects hardest to judge were those with a liking for "contemplative observation" and analysis of experience. Traits that were consistently well judged were "inhibition-impulsion, apathy-intensity, placidity-emotionality, and ascendence-submission."

288 **Etkin, William**, ed. *Social Behavior and Organization Among Vertebrates.* Chicago: University of Chicago Press, 1964. 307 pp. *D G T*

Includes chapters on cooperation and competition, courtship, and sexual behavior, and gives particular attention to body movement in theories of communication ("The Evolution of Signaling Devices," by Niko Tinbergen) and to types of social organization.

289 **Exline, Ralph V**. "Explorations in the Process of Person Perception: Visual Interaction in Relation to Competition, Sex, and Need for Affiliation." *Journal of Personality* 31 (1963):1-20. *D G T*

In this important initial study of gaze behavior, Exline presents evidence that women look at each other far more than men do; that competition inhibits mutual glances among individuals who in previous testing show a high need for "affiliation," but it increases mutual glances among "low affiliators." Groups of three women or three men, either high or low affiliators, were put into a competitive or noncompetitive experimental situation. The author discusses observation techniques, research instruments (the Esterline-Angus 20 Pen Operations Recorder), and the good observer reliability obtained in detail.

290 **Exline, Ralph V.; Gray, David; and Schuette, Dorothy**. "Visual Behavior in a Dyad as Affected by Interview Content and Sex of Respondent." *Journal of Personality and Social Psychology* 1 (1965):201-209. *D T*

Subjects were interviewed with either very personal questions or "innocuous" ones. Their gaze behavior was recorded and synchronized with incidents of speech. Results showed that women looked at the interviewer more than men, whether the interviewer was male or female, and that both men and women looked more at the interviewer during the "innocuous" questions than during the "embarrassing" questions.

291   **Exline, Ralph V., and Merrick, David.** "The Effects of Dependency and Social Reinforcement upon Visual Behaviour during an Interview." *British Journal of Social and Clinical Psychology* 6 (1967):256-266. *D T*

Male subjects scored as "dependent" or "dominant" from Schutz's FIRO-B tests were interviewed about leisure time and given high or low verbal reinforcement for eye contact. The research room and recording equipment are illustrated, and samples of the patterns of subjects' and experimenters' looking and speaking are presented. Significantly, dependent subjects given low reinforcement returned looks more than those given high reinforcement.

292   **Exline, Ralph; Thibaut, John; Brannon, Carole; and Gumpert, Peter.** "Visual Interaction in Relation to Machiavellianism and an Unethical Act." *American Psychologist* 16 (1961):396.

An abstract of a research report that subjects, particularly "low Machiavellian" ones, who had cheated later showed decreased "mutual visual interaction" with the interrogator.

293   **Exline, Ralph V., and Winters, Lewis C.** "Affective Relations and Mutual Glances in Dyads." In *Affect, Cognition, and Personality*, edited by S. S. Tomkins and C. E. Izard, pp. 319-350. New York: Springer Publishing Co. 1965. *D G T*

Two experiments testing the hypothesis that the degree of mutual glancing or eye contact is related to affective orientation, i.e., positive or negative feelings toward the other. In one experiment subjects were to discuss leisure time with an experimenter, who subsequently responded with neutral, positive, or negative feedback. Subjects who were given negative affect induction significantly decreased eye contact. A second experiment showed that, after subjects had stated which evaluator in a mock experiment they preferred, the women subjects exchanged more eye contact while speaking to the preferred evaluator. Men reduced eye contact much less when listening to the preferred evaluator.

294   **Fast, Julius.** *Body Language*. New York: M. Evans and Company, 1970. 192 pp. *N*

A popularized account of kinesics research.

295   **Faurbye, Arild.** "The Structural and Biochemical Basis of Movement Disorders in Treatment with Neuroleptic Drugs and in Extrapyramidal Diseases." *Comprehensive Psychiatry* 11 (1970):205-225. *T*

Written in medical and biochemical terms, this is a succinct and valuable review of research on specific motor disturbances (various types of rigidity, tremor, hypokinesia, bradykinesia, choreatic movements, dyskinesia, and stereotyped movements), with tentative conclusions about their structural and biochemical bases and a discussion of the neurological side effects of phenothiazines used in treatment of psychosis.

296   **Faurbye, A.; Rasch, P.-J.; Petersen, P. Bender; Brandborg, G.; and Pakkenberg, H.** "Neurological Symptoms in Pharmacotherapy of Psychoses." *Acta Psychiatrica Scandinavica* 40 (1964):10-27. *T*

In a discussion of the effects of phenothiazines on the body movement of psychotic patients, the authors present clinical evidence for three drug-induced neurological syndromes: acute dystonia (or spasmodic movements), rigidity, and tardive dyskinesia (or rhythmic, stereotyped movements). The nature and course of the syndromes, their treatment, and theories regarding underlying neurophysiological mechanisms are discussed.

297 **Fay, Temple.** "The Origin of Human Movement." *American Journal of Psychiatry* 111 (1955):644-652.

An overview of the evolution of upright locomotion and limb-trunk coordination from fish to man, analyzing the "comparative anatomy of vertebrates' nervous systems" and the origins and functions of progression patterns recapitulated in the neonate's "swimming" patterns, the infant's amphibianlike crawling, and so on. The purpose of this paper is to show that early reflexes which survive cortical damage can be used in rehabilitating the disabled: for example, the spastic hand may be "unlocked" by putting the patient in positions that elicit reflexes which have evolutionary origins.

298 **Feldenkrais, Moshé.** *Body and Mature Behaviour: A Study of Anxiety, Sex, Gravitation and Learning.* London: Routledge & Kegan Paul, 1949. 167 pp. D. [Reprint ed., New York: International Universities Press, 1966]

This treatise on neurophysiological, anatomical, and reflexive aspects of behavior and learning includes sections on antigravity mechanisms, body mechanics, and posture; the relation between kinesthetic sensations and vestibular reactions to orientation in space; the origins of anxiety in bodily reactions to falling; correct posture and balance; and motor aspects of sexual activity. Throughout there are references to psychological aspects of movement, posture, and muscle tension.

299 **Feldman, Paul E.** "Clinical Evaluation of Butaperazine." *Psychosomatics* 8 (1967):26-28. T

An example of research on the effects of phenothiazine drugs on hospitalized schizophrenic patients, with data on extrapyramidal effects and psychomotor behavior. In this case, this drug was considered efficacious for patients with "increased psychomotor manifestations."

300 **Feldman, Sandor.** "The Blessing of the Kohenites." *American Imago* 2 (1941):296-322.

A psychoanalytic interpretation of a Jewish ritual, the "raising up of hands" performed by the *kohen*, or priest. The rite is carefully described, and its relation to the Oedipal situation is analyzed.

301 **Feldman, Sandor S.** *Mannerisms of Speech and Gestures in Everyday Life.* New York: International Universities Press, 1959. 298 pp. [Latest edition: 1969]

Part Two of this book includes psychological interpretation of a wide range of movement: movements of the head, facial expression, sexual and aggressive aspects of touch and tickling, body expressions of pride, shoulder gestures, gaze interaction, psychological significance of blinking, gait, mannerisms of loosening the collar, ring play, smoking, spitting, yawning, fingering objects, and various gestures signifying honesty, contempt, etc. Some of the analyses include examples from psychoanalysis.

302 **Feleky, Antoinette M.** "The Expression of the Emotions." *Psychological Review* 21 (1914):33-41 *P T*

One hundred subjects were shown eighty-six posed and rather melodramatic photographs of different facial expressions and asked to indicate which of a long list of terms for emotion these expressed. The judges' interpretations are presented in table form, but no analysis of the results is made by the author.

303   **Fenichel, Otto.** *The Collected Papers of Otto Fenichel: First Series.* New York: W. W. Norton & Company, 1953. 408 pp. [Latest edition: 1963]

In Chapter 14 there is a noteworthy psychoanalytic discussion of the repressive function and libidinized aspects of muscle tension.

304   **Ferenczi, Sándor.** *Sex in Psycho-Analysis: Contributions to Psycho-Analysis.* Translated by E. Jones. Boston: Richard G. Badger, 1916. 338 pp.

In the following sections Ferenczi analyzes body expressions: the interpretation of transitory "expression displacements," body sensations, and muscular tensions during the analytic hour (pp. 193-212); "magic gestures" and the omnipotence period of early infancy, with a note about the regressive significance of epileptic attacks (pp. 220-235); and erotic aspects of catatonic symptoms (pp. 282-295).

305   **Ferenczi, Sándor.** "Psycho-analytical Observations on Tic." *International Journal of Psycho-Analysis* 2 (1921):1-30.

"Tic" in this paper refers to various stereotyped movements, from twitches to compulsive scratching. Drawing on the observations of Meige and Feindel in a 1903 book entitled *Tic, Its Nature and Treatment,* Ferenczi discusses the etiology of tics, theorizing that they are based on a tendency to abreact, to ward off suffering, and to onanism or erotic sensation transferred to another part of the body. He discusses the character of the tiquer, the similarities of tics to catatonic symptoms, and the differences between tics and obsessive actions or hysterical conversion reactions.

306   **Ferenczi, Sándor.** *Further Contributions to the Theory and Technique of Psycho-Analysis.* Translated by J. I. Suttie. London: The Hogarth Press, 1926. 473. pp.

In this collection there are papers on bodily symptoms at the end of an analytic session (pp. 239-241); the relation between muscle innervation, thinking, and attention (pp. 230-232); the effects of active intervention by the analyst and enactment by the patient in sessions (pp. 198-217); and the libidinal basis of tics, catatonic rigidities, and obsessive ceremonials (pp. 142-174).

307   **Fields, Sidney J.** "Discrimination of Facial Expression and Its Relation to Personal Adjustment." *American Psychologist* 5 (1950):309.

A summary of a report that subjects' ability to discriminate facial expression in a controlled experiment correlated with "social adjustment" as measured.

308   **Fine, Leon J.** "Nonverbal Aspects of Psychodrama." In *Progress in Psychotherapy,* vol. 4, edited by J. H. Masserman and J. L. Moreno, pp. 212-218. New York: Grune & Stratton, 1959.

The author describes ways in which concentrating on nonverbal behavior facilitates group interaction, training of psychodrama staff, and a clearer understanding of group process.

309   **Fine, Reiko; Daly, Dennis; and Fine, Leon.** "Psychodance: An Experiment in Psychotherapy and Training." *Group Psychotherapy* 15 (1962):203-223.

An excellent presentation of the therapeutic use of dance and movement in the treatment of chronic schizophrenic patients. The authors describe a program involving female patients, staff, and a psychodance leader which begins with dance and gradually evolves into psychodrama as the patients become more active and involved. The article includes theoretical assumptions about the significance of nonverbal behavior and the value of psychodance for both patients and staff, together with illustrations of psychodance techniques from events occurring within sessions.

310 **Finzi, Hilda.** "Interchangeability of Perceptual and Motor Activity in the Release of Organismic Tension." Ph.D. dissertation, New York University, 1961. (Datrix order no. 62-1497)

Forty-eight subjects were given Barron M Threshold Test inkblots to examine the increase or decrease in M (apparent movement) responses (considered one outlet for organismic tension) after four conditions: (1) acting out charades, (2) doing a repetitive exercise, (3) answering questions requiring creative thinking, and (4) standing immobile at attention. The results showed that inhibition of mobility (4) was followed by an increase in M perceptions; expressive movement (1) was followed by decrease in M; repetitive, meaningless movement (2) led to an increase of M and vocalizations of resentment; and creative thinking (3) did not reduce the perception of M on the inkblots.

311 **Fischer, Liselotte K.** "The Significance of Atypical Postural and Grasping Behavior during the First Year of Life." *American Journal of Orthopsychiatry* 28 (1958):368-375. *T*

Noting that, in Central African countries where infants are in constant physical contact with adults, they develop gross motor patterns much faster than Western children, whereas at the other extreme Spitz's institutionalized infants, deprived of contact, withdrew into "pathognomonic positions" and death, the author calls for more direct study of institutionalized infants to study the effects of environment on development and the distinctions between organic and psychogenic factors in motor abnormalities. Studies are reported showing that far fewer three-month-old, institution-reared infants were able to stand briefly with support than were family-reared infants and that abnormal hoarding and grasping behavior was common in the institutionalized children.

312 **Fisher, Seymour, and Cleveland, Sidney E.** *Body Image and Personality.* (D. Van Nostrand Company, 1958; reprint ed., New York: Dover Publications, 1968). 448 pp. *T*

Perhaps the most extensive work on body image, this book includes an excellent bibliography and review of the literature. Included are sections on the boundary dimension and "barrier score" in relation to psychosomatic phenomena; body image and personality traits such as suggestibility and independence; and body image and values. There are chapters dealing with case studies of "high-barrier" and "low-barrier" men; body image boundaries and small-group behavior; body image alterations in psychoses and neuroses; and boundaries and physical reactivity. Additional parts deal with family patterns and sex and cultural differences in body image boundaries. Although the authors focus on their own theories and research, there is continual inclusion of other research and literature on body image.

313 **Fleishman, Edwin A.** "Dimensional Analysis of Movement Reactions." *Journal of Experimental Psychology* 55 (1958):438-453. *D T*

An example of research on individual differences in motor performance done in such areas as airplane pilot training, this study is relevant here for two reasons. First, it illustrates and describes twenty different pieces of equipment for measuring complex motor patterns, which may be useful in a number of research areas. Second, it reports on results of a factor analysis of a battery of motor tests yielding reliabilities and intercorrelations and some seven psychologically meaningful factors of complex motor performance.

314 Fleishman, Edwin A., and Ellison, Gaylord D. "A Factor Analysis of Fine Manipulative Tests." *Journal of Applied Psychology* 46 (1962):96-105. *D T*

Seven hundred and sixty Air Force trainees were given twenty-one motor tests of hand skill, some of them paper tests, some using apparatus. The tests are described and illustrated. The results of a factor analysis indicate five factors: manual dexterity, finger dexterity, speed of arm movement, wrist-finger speed, and aiming.

315 Foley, John P. "Judgment of Facial Expression of Emotion in the Chimpanzee." *Journal of Social Psychology* 6 (1935):31-67. *P T*

This paper has an excellent bibliography and review of research on human facial expression from the 1600's to 1935, including European authors not reviewed for the present bibliography. Foley also reports research on observer judgment of six photographs of a chimpanzee showing various emotions. There was little agreement, and when judgments were compared with the context in which the expression occurred and presumably the real emotion, there were dramatic contradictions, including interpreting the animal's expression of rage as laughter.

316 Foss, B. M., ed. *Determinants of Infant Behaviour.* London: Methuen & Co.; New York: John Wiley & Sons, 1961. 308 pp. *G P T*

This is the first in an excellent series on infant behavior and mother-infant interaction. The following chapters include a great deal of information on movement patterns: a film study of neonate response to stimulation, by Helen Blauvelt and Joseph McKenna; infant behavior at the breast, by Mavis Gunther; abnormal movements in breech babies, by H. F. R. Prechtl; feeding patterns in kittens, by Jay S. Rosenblatt, Gerald Turkewitz, and T. C. Schneirla; infant monkey responses to cloth and wire surrogate mothers and to various forms of isolation, by Harry F. Harlow; treatment of a disturbed two-year-old child and a study of nurse-infant interaction, by Geneviéve Appell and Myriam David; differences in the social behavior of institution versus home-raised infants, by Harriet L. Rheingold; and the development of the smiling response in early infancy, by J. A. Ambrose.

317 Foss, B. M., ed. *Determinants of Infant Behaviour.* Vol. 2. London: Methuen & Co.; New York: John Wiley & Sons, 1963. 248 pp. *G P T*

These are the proceedings of the Second Tavistock Seminar on Mother-Infant Interaction. The following sections include extensive movement observations: how the mother monkey relates to her infant, and responses of infants with different types of mothering, by Harry F. Harlow; expressive and social behaviors in infant monkeys, by Thelma E. Rowell; the mother-child interaction in babies with minimal brain damage, by H. F. R. Prechtl; the development of infant-mother interaction among the Ganda, by Mary Ainsworth; early development of smiling and patterns of interaction within a family, by Peter H. Wolff; conditioning infant exploratory behavior, by Harriet L. Rheingold; critical periods and development of smiling, by J. A. Ambrose; and the nature of imprinting, by R. A. Hinde.

318 Foss, B. M., ed. *Determinants of Infant Behaviour.* Vol. 3. London: Methuen & Co.; New York: John Wiley & Sons, 1965. 264 pp. *G P T*

This collection includes a study of synchrony between the maternal behavior cycle and baby rat development, by Jay S. Rosenblatt; research methodology and monkey mother-infant interaction, by Gordon D. Jensen and Ruth A. Bobbitt; "Rhesus Monkey Aunts," by R. A. Hinde; behavior of a human-reared baboon, by Thelma Rowell; how various animal and human babies are carried, by Harriet L. Rheingold and Geraldine C. Keene; case studies of mother-infant interaction, by Joyce Robertson; mother-child interaction at thirteen months, by Geneviéve Appell and Myriam David; a study of social behavior in twenty pairs of infant twins, by D. Freedman; social behavior differences among youngest and only Israeli infants, by J. L. and Hava B. Gewirtz; imitation and identification, by B. M. Foss; and the development of smiling in Israeli infants from four different environments by J. L. Gewirtz. As with the previous proceeding in the series, group discussion follows each paper.

319   **Foss, B. M.,** ed. *Determinants of Infant Behaviour.* Vol. 4. London: Methuen & Co., 1969. 304 pp. *G P T*

This collection maintains the caliber of previous papers by the Tavistock Study Group on Mother-Infant Interaction and includes a number of articles with movement observations: a review of mother-infant interaction observations from primate field studies, by David A. Hamburg; a follow-up report of isolated or deprived monkeys, by Harry F. and Margaret K. Harlow; the influence of temporary separation on infant monkeys, by R. A. Hinde; modes of decreasing the mother-infant bond in Macaque monkeys, by I. Charles Kaufman and Leonard A. Rosenblum; three studies of infant reaction to strangers or to a strange situation, by Mary D. Salter Ainsworth and Barbara A. Wittig, Harriet L. Rheingold and George A. Morgan, and Henry N. Ricciuti; intelligence, motor capacities, and social behavior of thalidomide children, by Thérèse Gouin Décarie; a longitudinal study of mother-child interaction, by Louis W. Sander; and social behavior differences among Israeli children reared in four different environments, by Hava B. and Jacob L. Gewirtz.

320   **Francis, Robert J., and Rarick, G. Lawrence.** "Motor Characteristics of the Mentally Retarded." Unpublished paper, University of Wisconsin, 1957. 86 pp. *G T* (There are copies at the Library of Congress)

Primarily a study of the motor performance of educable but retarded schoolchildren, as compared with that of normal children, on tests of strength, balance, and agility. A small sample group of institutionalized retarded children were also studied. The tests—which included running, jumping, balance beam, grip strength —are described. Age and sex differences are noted, and marked superiority of the normal children on the tests is reported. The performance of the institutionalized children compared with that of three- to four-year-old normal children.

321   **Francis-Williams, Jessie, and Yule, William.** "The Bayley Infant Scales of Mental and Motor Development: An Exploratory Study with an English Sample." *Developmental Medicine and Child Neurology* 9 (1967):391-401. *G T*

A study of 300 English infants, ages one to fifteen months, in which specific differences in motor development between English and American infants are described.

322   **Frank, Lawrence K.** "Tactile Communication." *Genetic Psychology Monographs* 56 (1957):209-255.

In one of the most extensive theoretical discussions of this neglected topic, the author discusses physiological, developmental, cultural, and psychopathological aspects of touch from a "transactional" point of view. He also refers to some of the relevant literature and indicates areas for research.

323 **Frazer, Sir James G.** *The Golden Bough: A Study in Magic and Religion.* (Macmillan & Co., 1922; reprint ed., New York: The Macmillan Company, 1951). 864 pp.

This classic work on myths, rites, and religions of different eras and cultures is replete with descriptions of dances, rituals, taboo acts, and magic practices and their function in society.

324 **Freedman, Norbert.** "The Analysis of Movement Behavior during the Clinical Interview." Unpublished paper, Downstate Medical Center, n.d. 33 pp. G (Copies from author; Clinical Behavior Research Unit, Downstate Medical Center, Box #88, 450 Clarkson Avenue, Brooklyn, N.Y. 11203)

The number and duration of "object-focused" and "body-focused" movements of five paranoid and five depressed patients occurring in a ten-minute sample of initial and follow-up interviews were studied. Paranoid patients had more object-related movements than the depressed patients, and in follow-up the number decreased. Various types of object- and body-focused movements are described (e.g., "speech primacy" and "motor primacy"); the motor-primacy movement decreased markedly with improved symptomatology. Certain types of body-focused acts were more prevalent in the depressed group. Fluctuations of the activity of one patient during an interview are analyzed and compared with speech content, major pauses, and volume. Psychopathological aspects, degree of "communicative intent," and various types of body- or object-focused movements are analyzed in a theoretical discussion.

325 **Freedman, Norbert.** "Body Movements and the Verbal Encoding of Aggressive Affect." Unpublished paper, Downstate Medical Center, 1971. 37 pp. T (Copies from author, Clinical Behavior Research Unit, Downstate Medical Center, Box #88, 450 Clarkson Avenue, Brooklyn, N.Y. 11203)

Four types of object-focused movements and three kinds of body-focused movements are defined. From five-minute video samples, the movements of twenty-four women subjects were studied and various types were compared with types of hostility, as determined from the verbal content. Subjects with a high degree of certain object-focused movements expressed aggression overtly; those with hand-to-hand body-focused activity expressed hostility covertly. An analysis of patterns within the five minutes suggested that the movement may facilitate a "buildup" to hostility or soften it once it is expressed.

326 **Freedman, Norbert, and Hoffman, Stanley P.** "Kinetic Behavior in Altered Clinical States: Approach to Objective Analysis of Motor Behavior during Clinical Interviews." *Perceptual and Motor Skills* 24 (1967):527-539. T

Videotape analysis of the hand movements of paranoid patients in psychotherapy yielded two major distinctions: body-focused and object-focused referring roughly to movements in toward the body or touching the body, chair, etc., and movements away from the body, as in speech gesticulation). Five types of object-focused movements—punctuating, minor qualifiers, literal-productive, literal-concretization, and major qualifiers (gross, chaotic, "autonomous" movements)—are defined as on a continuum from an integral speech-movement relationship to an autonomy of the motor sphere. Two case presentations show how the movement changed from acute psychotic episodes to less acute phases. One patient, described later as depressed, showed more body-focused movement; the other, who was at first very disorganized, later showed an increase in speech-movement integration.

327   **Freedman, Norbert; O'Hanlon, James; Oltman, Philip; and Witkin, Herman A.** "The Imprint of Psychological Differentiation on Kinetic Behavior in Varying Communicative Contexts." Unpublished paper, Downstate Medical Center, 1971. 57 pp. *G T* (Copies from N. Freedman, Clinical Behavior Research Unit, Downstate Medical Center, Box #88, 450 Clarkson Avenue, Brooklyn, N.Y. 11203)

To study the relationship between body- and object-focused movements and both differing interpersonal contexts and levels of psychological differentiation, twelve field-dependent and twelve field-independent college women were videotaped in interviews with "warm" and "cold" interviewers. Hand-to-hand activity and "motor primacy" gestures indicated limited differentiation. However, certain changes in kinetic behavior were related to a shift in interviewers, notably a reduction of object-focused activity with the "cold" interviewer and persistence of body touching following a "cold" encounter.

328   **Freeman, G. L.** "The Facilitative and Inhibitory Effects of Muscular Tension upon Performance." *American Journal of Psychology* 45 (1933):17-52 *P T*

A series of experiments investigating the relationship between muscle tension and anticipation, task difficulty, reaction time, reaction to shock, and facilitation of performance are reported. An apparatus for photographing "tendon deformation" under tension was used.

329   **Freeman, G. L.** "Postural Tensions and the Conflict Situation." *Psychological Review* 46 (1939):226-240.

In this theoretical paper on postural tensions as a reflection of thwarted "activity-in-progress" and conflict between directed action and inhibition, the author cites experimental evidence, including his own extensive research on psychological aspects of muscle tension.

330   **Freud, Sigmund.** "Symptomatic and Chance Actions." In *The Basic Writings of Sigmund Freud*, translated and edited by A. A. Brill, pp. 129-140. New York: Random House (Modern Library), 1938.

Freud describes incidents of fidgeting, chance actions, loss of objects, etc., observed in psychoanalytic sessions and everyday life. He interprets these, when considered within the person's history and current situation, as symptomatic or symbolic expressions of unconscious conflicts and affects.

331   **Freud, Sigmund.** "Infantile Sexuality." In *The Basic Writings of Sigmund Freud*, translated and edited by A. A. Brill, pp. 580-603. New York: Random House (Modern Library), 1938.

Within this chapter thumb-sucking, masturbation, and muscular activity are discussed as manifestations of infantile sexuality and analyzed as precursors of adult sexuality and sources of pleasure. Particular note is made of the function of rhythms in such activities.

332   **Freud, Sigmund.** "Obsessive Acts and Religious Practices." In *Collected Papers*, edited by E. Jones, pp. 25-35. New York: Basic Books, 1959.

The etiology and defensive function of obsessive acts and everyday "ceremonials."

333   **Freud, Sigmund.** *Jokes and Their Relation to the Unconscious.* Translated and edited by J. Strachey. New York: W. W. Norton & Company, 1960. 258 pp.

In one passage (pp. 189-198), Freud analyses the origins of comic pleasure in empathy, particularly from a process by which one compares one's own movements with the exaggerated movements of the clown. In this discussion he remarks on the relationship between attention and motor functions and how "ideational mimetics" and motor innervation accompany and reflect thoughts as well as emotions.

334 **Freud, Sigmund**. "General Remarks on Hysterical Attacks (1909)." In *Dora: An Analysis of a Case of Hysteria*, edited by P. Rieff, pp. 153-157. New York: The Macmillan Company (Collier Books), 1963.

A concise summary of Freud's theories about the complex origins and intra-psychic processes inherent in hysterical attacks, which he calls "nothing but fantasies projected and translated into motor activity and represented in pantomime."

335 **Fries, Margaret, and Lewi, Beatrice.** "Interrelated Factors in Development: A Study of Pregnancy, Labor, Delivery, Lying-In Period and Childhood." *American Journal of Orthopsychiatry* 8 (1938):726-752. *T*

One of the first papers in which Dr. Fries discusses her research on "activity types," assessed through observing the amount and kind of movement response to stimuli and observed to be individually consistent over time. This paper discusses three children in detail, the mother-child interaction and predictions based on their health, physical activity, environment, the emotional health of the parents, etc. The article is particularly notable here because charts used to assess the infants, including the amount of movement of different parts, etc., are reprinted.

336 **Fries, Margaret E., and Woolf, Paul J.** "Some Hypotheses on the Role of the Congenital Activity Type in Personality and Development." In *Psychoanalytic Study of the Child,* vol. 8, pp. 48-62. New York: International Universities Press, 1953.

The "congenital activity type," based on the amount and kind of motor activity of neonates in terms of hypoactive, quiet, moderately active, active, and hyper-active, is discussed in terms of its implications for personality formation, parent-child relations, psychosexual and ego development, and later defense mechanisms. Some cases are presented in a section on the activity type and a predisposition to psychopathology.

337 **Frijda, Nico H.** "The Understanding of Facial Expression of Emotion." *Acta Psychologica* 9 (1953):294-362. *G P*

Subjects observed films or photographs of individuals reacting to various stimuli and recorded their judgments of the emotions expressed, the situations, etc. Evaluation of the "correctness" of the judgments indicated that there are invariable meanings in different aspects of facial expression and that responses to film examples are superior to judgments of photographs. The role of knowledge of the situation in supporting the interpretations and an analysis of the judges' impression formation is presented. Various aspects of the "expression melody," or pattern, are defined and analyzed with respect to the judgments. One of the most sophisticated, complex, and sensible studies of facial expression recognition.

338 **Frijda, Nico H.** "Facial Expression and Situational Cues." *Journal of Abnormal and Social Psychology* 57 (1958):149-154. *P T*

Two groups were shown photographs of different facial expressions: one group was given one set of descriptions of the situation accompanying each expression, and the other group was given a different set. The subjects' interpretations of the emotions expressed were compared, and showed that there were differences due to situational cues but also similarities due to recognition of what the author calls "general attitudes," which are the "intrinsic meaning of facial expression."

339   **Frijda, Nico H.** "Facial Expression and Situational Cues: A Control." *Acta Psychologica* 18 (1961):239-244. *T*

To test the possibility that similarities in interpretation found in a previous experiment might be a result of situational cues alone, and not facial cues, the author added a control group that was given only the descriptions of the situation. The results confirmed the author's hypothesis that facial expression indicates "a general aspect of the person's emotional state," whereas the situational cues specify it.

340   **Frijda, Nico H.** "Recognition of Emotion." In *Advances in Experimental Social Psychology*, vol. 4, edited by L. Berkowitz, pp. 167-223. New York: Academic Press, 1969. *D T*

With an up-to-date review of research on recognition of emotion from facial expression, the author outlines the methodological and theoretical issues involved. He reports his own extensive research on judgment of emotion and motivational state from photographs and films with respect to dimensional theories of emotion, situational cues, cue conflict situations, and the complexity of the recognition process.

341   **Frijda, Nico H., and Van De Geer, John P.** "Codability and Recognition: An Experiment with Facial Expressions." *Acta Psychologica* 18 (1961):360-367. *T*

"Identifiability measures" obtained from judgments of facial expressions were compared with "recognition scores" obtained from a procedure in which the subjects found photographs judged in the first procedure mixed among a large number of photographs. Regression analysis of the relationship between the two scores was made. The results indicate that, if there is little agreement of interpretation, there will also be poorer recognition of the expression among a group of photographs.

342   **Frings, Hubert and Mabel.** *Animal Communication*. Waltham, Mass.: Blaisdell Publishing Company (Ginn and Company), 1964. 204 pp. *D P*

This book for the nonspecialist or layman presents a wealth of information on the body movements and other communication mechanisms of various animals in courtship, "social cooperation," and care for the young, plus a section on methods of study and research equipment.

343   **Frisch, Karl von.** *The Dance Language and Orientation of Bees*. Cambridge, Mass.: Harvard University Press (Belknap Press), 1967. 566 pp. *D G P T*

An exhaustive analysis of the behavior of the bee, particularly of its dances as communication. Observation methods, equipment, and results are presented and illustrated with great detail. The bee varies its incredible dance to communicate distance, direction, and the value of a goal to its comrades, who then find it without being led.

344   **Frois-Wittman, J.** "The Judgment of Facial Expression." *Journal of Experimental Psychology* 13 (1930):113-151. *P T*

A comprehensive study of observer judgments of facial expression and the muscular basis for them, using posed photographs and drawings of all or part of a face. In the research, 165 subjects judged up to 227 pictures and checked emotions from a list of forty-three terms. Later analysis of thirty-two "modal" expressions indicated that there were distinct patterns of expression and muscular involvement for a given emotion. The author found agreement on some terms and high "secondary" frequency on others.

345   Fulcher, John S. "'Voluntary' Facial Expression in Blind and Seeing Children." *Archives of Psychology*, no. 272, 38 (1942):1-49. *G*

A group of 118 sighted children, ages four to sixteen, and a group of 50 blind persons, ages six to twenty-one, served as subjects in an experiment in which they were asked to perform four different facial expressions. Photographs of their happy, sad, angry, or frightened expressions were analyzed and judged for "adequacy." The author reports a number of similarities and differences between the blind and seeing children, younger and older subjects, and boys and girls. For example, although the blind and seeing children show similar expressions, those of the seeing children are more differentiated and increase in activity with age. He discusses possible ways that the blind children may acquire the expressions they have.

346   Gage, N. L. "Judging Interests from Expressive Behavior." *Psychological Monographs*, Whole No. 350, 62 (1952):1-20. *G*

Four different groups of judges observed strangers going through a number of tasks (drawing, lighting and holding a match, building a house of cards, patty-caking, and describing the room) and then filled out Kuder Preference Records of individual interests the way that they imagined the stranger would. Their predictions correlated well with the actual responses of the strangers, but there was evidence that clues of subculture stereotypes influenced the predictions more than expressive behavior did.

347   Gallini, Giovanni Andrea. *A Treatise on the Art of Dancing.* (1762; reprint ed., New York: Broude Brothers, 1967). 292 pp. *D*

An eighteenth-century treatise on dance and dance aesthetics, including English dances of the time and classic dance; notes on dances of Europe, Asia, and Africa; and a chapter on mime.

348   Garfiel, Evelyn. "The Measurement of Motor Ability." *Archives of Psychology* 62 (1923):1-47. *D T*

A series of structured motor tests were given to fifty college women to determine whether there is a general motor ability and, if so, how it relates to intelligence. The results, as well as those cited in a good review of other tests and related research, indicate that there is a general motor ability but that it is independent of I.Q.

349   Garfield, John C. "Motor Impersistence in Normal and Brain-Damaged Children." Ph.D. dissertation, State University of Iowa, 1963. 86 pp. (Datrix order no 64-3371)

"Motor impersistence," or the inability to sustain simple motor acts, was measured in 140 normal children, ages five to eleven, and in 25 nonretarded but brain-damaged children of the same age range. Of these, 68 percent of the brain-damaged children showed impersistence on two or more tasks (e.g., keep eyes closed), as compared with 3 percent of the normal group.

350   Garfield, John C.; Benton, A. L.; and MacQueen, J. C. "Motor Impersistence in Brain-Damaged and Cultural-Familial Defectives." *Journal of Nervous and Mental Disease* 142 (1966):434-440. *T*

After reviewing research on the phenomenon of "motor impersistence," the inability to sustain a motor act, in brain-damaged individuals, the authors investigate whether retardates who are clearly brain-damaged have motor impersistence unlike cultural-familial retardates. Forty-four children and adults with mild retardation (I.Q.'s of 55 to 84) performed simple tasks like keeping the eyes closed or fixing their gaze. The brain-damaged retardates showed motor impersistence to a far greater extent, and younger subjects of both groups tended to be more impersistent.

351   **Gaskin, L. J. P.**, compiler. *A Select Bibliography of Music in Africa.* London: International African Institute, 1965. 83 pp.

A bibliography that has a geographically arranged section on African dance literature and a list of bibliographies on African music and dance.

352   **Gates, Alice.** *A New Look at Movement: A Dancer's View.* Minneapolis, Minn.: Burgess Publishing Company, 1968. 187 pp. *D*

An analysis of body movement influenced by the work of Rudolf Laban and a number of American modern dancers, designed primarily for dance and physical education. There are chapters on how to observe movement and what to look for in dance and everyday activity; problems of becoming aware of one's own movement; analysis of spatial patterns, initiation, force and weight factors, rhythm, etc.; classification of movement; interpretation of meaning and expression; and implications for dance aesthetics and dance education.

353   **Gates, Georgina S.** "A Test for the Ability to Interpret Facial Expression." *Psychological Bulletin* 22 (1925):120.

A brief summary of a conference report on children's judgment of emotions from photographs (ages three to fourteen). It was possible to gauge a child's ability to interpret. This ability increased with age and correlated more with social maturity than with mental maturity.

354   **Geber, Marcelle, and Dean, R. F. A.** "Gesell Tests on African Children." *Pediatrics* 20 (1957):1055-1065. *G*

Gesell development tests were given to 183 children from age one month to six years in a number of locations in Uganda, East Africa, which differed in social class and condition. The report is largely descriptive, with only rough quantitative results in terms of the average scores relative to American and European norms. Another study of 113 neonates is reported reiterating the results of the first, namely, that these African infants are very precocious in motor development, intellectual ability, and social behavior in the first year of life, with their motor skills remarkably advanced compared with European infants. Some aspects of child raising are noted to account for this precocity (e.g., constant handling of the infant). Changes in developmental quotients and less progress after age two are then described, with numerous examples given of possible environmental influences.

355   **Gerard, Margaret W.** "The Psychogenic Tic in Ego Development." *Psychoanalytic Study of the Child* 2 (1946):133-162.

The history and treatment of four children with severe tics are described in detail. There is evidence that in each case the tic symptom was related to a traumatic experience and a history of inhibition of aggression.

356   **Gesell, Arnold.** "Reciprocal Interweaving in Neuromotor Development: A Principle of Spiral Organization Shown in the Patterning of Infant Behavior." *Journal of Comparative Neurology* 70 (1939):161-180. *D G P T*

A theoretical paper with illustrations proposing that there is a pattern of alternation and progressive reintegration, "a developmental fluctuation of dominance of flexors versus extensors and also in unilateral and crossed lateral versus bilateral muscle groups" (p. 161), which serves development of uprightness and prehension in the infant. Charts of the stages of alternating dominance of flexion and extension and graphs of unilateral-bilateral patterns in the first year of life are presented.

357   **Gesell, Arnold, and Amatruda, Catherine S.** *Developmental Diagnosis: Normal and Abnormal Child Development Clinical Methods and Pediatric Applications.* 2d rev. ed. New York: Paul B. Hoeber, 1947. 496 pp. *D G T*

This book contains a wealth of information about normal developmental stages of motor behavior, its assessment from birth to age three; and the movement factors in retardation, endocrine, convulsive, and neurological disorders and brain damage. In addition, there are notes on the motor and expressive characteristics of blind, deaf, premature, precocious, institutionalized, and adopted children. There are excellent illustrations of the movement patterns of various motor disorders and appendixes describing diagnostic procedures, inventory forms, and examination equipment.

358   **Gesell, Arnold, with the assistance of Louise Bates Ames.** "Early Evidences of Individuality in the Human Infant." *Scientific Monthly* 45 (1937):217-226. *G P T*

Film analysis of the behavior traits of a few infants were compared with a follow-up appraisal at five years of age. "Energy output, motor demeanor, self-dependence, emotional expressiveness, and readiness of smiling" were persistent traits. Analysis of consistencies in "motor habitude" is also presented.

359   **Gesell, Arnold, and Ames, Louise Bates.** "The Development of Handedness." *Journal of Genetic Psychology* 70 (1947):155-175. *G T*

The results of an extensive film study of patterns of hand dominance and bilaterality-unilaterality in children from eight weeks to ten years old in structured situations. Handedness goes through a number of changes and alternations, which are summarized here, and marked individual differences are noted.

360   **Gesell, Arnold, and Halverson, Henry M.** "The Daily Maturation of Infant Behavior: A Cinema Study of Postures, Movements, and Laterality." *Journal of Genetic Psychology* 61 (1942):3-32. *D G T*

One infant was filmed for brief intervals every day from age 15 to 235 days in a controlled situation within her home. Detailed analysis of changes in eye and head movement and arm and leg behavior are presented, often in terms of the percentage of time in which a position was maintained: e.g., right head aversion prominent up to 82 days. The kinds of arm movements observed at each of fourteen stages in terms of laterality and flexion-extension; the incidence of hand to mouth, thumb-sucking, and fingering of abdomen; the patterns of laterality of postures; prehensile development; and leg movements in terms of predominance of flexion or extension are described.

361   **Gesell, Arnold, and Ilg, Frances L.** *Feeding Behavior of Infants: A Pediatric Approach to the Mental Hygiene of Early Life.* Philadelphia: J. B. Lippincott Company, 1937. 201 pp. *G P T*

Through film analysis and naturalistic observation of infants from birth, every facet of feeding behavior is examined: oral reflexes; suckling movements; head, hand, and body activity; expressive behavior; use of feeding implements; developmental stages in feeding behavior; and the mother-infant relationship. The book is extensively illustrated with photographs and charts.

362 **Gesell, Arnold, and Thompson, Helen.** "Learning and Growth in Identical Infant Twins: An Experimental Study by the Method of Co-Twin Control." *Genetic Psychology Monographs* 6 (1929):1-124. *D G P T*

Twin girls were studied from one to eighteen months of age, and for a six-week period up to age fifty-two weeks one twin was trained in stair climbing and cube manipulation but the other was not. Extensive analysis of their motor skills was done partly from films. Notably, at fifty-three weeks the untrained twin spontaneously and easily performed the tasks, thus indicating that maturation was more influential than training. Charts of their motor and visual development, detailed descriptions of how they performed tasks and moved (with remarkable similarities); and comparisons of their expressive and gestural behavior are presented.

363 **Gesell, Arnold, and Thompson, Helen.** "Twins T and C from Infancy to Adolescence: A Biogenetic Study of Individual Differences by the Method of Co-Twin Control." *Genetic Psychology Monographs* 24 (1941):3-121. *D G T*

In this important study one section is devoted to motor behavior, with detailed comparisons of the twins' motor development, facial expressions, posture, handwriting behavior and samples, dancing, laterality, timing in locomotion and climbing, and prehension. Together with other sections on physical development, language behavior, personal-social behavior, and the discussion of personality and environmental influences, the significance of the movement patterns observed can be studied relative to other aspects of behavior, personality, and development.

364 **Gibson, James J., and Pick, Anne D.** "Perception of Another Person's Looking Behavior." *American Journal of Psychology* 76 (1963):386-394. *D G*

Following a very interesting discussion about the ability to discriminate different "looks" and the effects of being looked at, the authors discuss a study of accuracy in judging whether one is being looked at. Subjects judged whether an experimenter who used three head positions and seven points of eye fixation was looking at them. Good discrimination and accuracy, somewhat affected by head position, was found.

365 **Giedt, F. Harold.** "Cues Associated with Accurate and Inaccurate Interview Impressions." *Psychiatry* 21 (1958):405-409.

Forty-eight psychotherapists rated personality traits of four patients from silent films, written protocols, and sound recordings and sound films of psychiatric interviews, and these ratings were compared with those of judges who had intensive contact with the patients. There was good agreement on seven traits (e.g., intelligence, anxiety, adherence to reality, etc.), and the cues that appeared most to influence the judgments are summarized in this paper: e.g., content, direct statements from the patient, and specific nonverbal characteristics for specific psychodynamics when words were not heard.

366 **Ginsburg, Ethel L.** "Three Studies in Hyperactivity: II, The Relation of Parental Attitudes to Variations in Hyperactivity." *Smith College Studies in Social Work* 4 (1933):27-53. *T*

A study in which a number of cases are presented providing evidence that the hyperactive behavior of twenty-five children was related to family conflicts and tensions.

367   **Glanville, A. Douglas, and Kreezer, George.** "The Characteristics of Gait of Normal Male Adults." *Journal of Experimental Psychology* 21 (1937):277-300. *D G T*

From the result of detailed film analysis of the gaits of ten men, measurements of many aspects of the walking pattern are presented.

368   **Goffman, Erving.** *The Presentation of Self in Everyday Life.* New York: Doubleday & Company (Anchor Books), 1959. 259 pp.

Although there are only brief references to expressive movement per se (notably pp. 24-25, 33-34, 51-53, 73-76, and 187), Goffman's exposition of behavior in social situations, using the model of role playing from the theater, puts movement behavior in an important theoretical perspective. He speaks specifically of the significance of setting, appearance, and manner, and in discussing manner he posits the function of ways of walking, gesturing, working, etc., in a social context as an important facet of one's "presentation of self."

369   **Goffman, Erving.** *Encounters: Two Studies in the Sociology of Interaction.* Indianapolis, Ind.: The Bobbs-Merrill Company, 1961. 152 pp.

Important theoretical writing on the nature of "focused interaction" in a gathering or "situated activity system," the rules for its regulation, and the role performances that characterize it. Body movement and nonverbal behavior is only briefly discussed or alluded to. As with other Goffman works, he sets the stage for later research in communication and interaction, much of it including movement, and defines terms and concepts in such a way as to directly facilitate this research.

370   **Goffman, Erving.** *Behavior in Public Places: Notes on the Social Organization of Gatherings.* New York: The Free Press, 1963. 248 pp.

An analysis of the nature and function of face-to-face social interaction, including sections on the "body idiom" (movements, appearance, sound level) in situations without spoken communication; types of social involvement and rules for their expression; and analysis of "focused interaction" and the behaviors that initiate, regulate, and terminate engagements among acquainted and unacquainted people. Examples from behavior in institutions, public places, and social gatherings frequently include references to movement, posture, or visual behavior.

371   **Golani, Ilan, and Zeidel, Shmuel** (notator). *The Golden Jackal.* Tel Aviv, Israel: Movement Notation Society, 1969. 124 pp. *D N P* (Copies from the Movement Notation Society, 75 Arlozorov Street, Holon, Israel)

A rare collaboration between a movement notator and a zoologist, this book is interesting as an example of how movement notation can be applied to animal studies. The behavior of a pair of jackals is recorded with Eshkol-Wachmann movement notation.

372   **Goldstein, Iris B.** "Role of Muscle Tension in Personality Theory." *Psychological Bulletin* 61 (1964):413-425.

A valuable review and bibliography of research on muscle tension and individual differences, reactions to stress, anxiety, psychodiagnosis, and specific physical or neurotic symptoms, with generalizations from the findings and a critique of the variety of measurements used.

373   **Goldstein, Kurt.** *The Organism: A Holistic Approach to Biology Derived from Pathological Data in Man.* New York: American Book Company, 1939. 533 pp. [Boston, Beacon Press, 1963]

A most valuable source for the understanding of movement behavior, particularly for its Gestalt approach to the mind-body controversy, its analysis of the nature of reflexes, and the holistic interpretation of "preferred behavior" and adaptation to impairment.

374 **Golla, F. L., and Antonovitch, S.** "The Relation of Muscle Tonus and the Patellar Reflex to Mental Work." *Journal of Mental Science* 75 (1929):234-241. D

Records from an optical myography apparatus for measuring foot and hand tension while the limbs were at rest show that during math calulations, silent reading, etc., the tonus rises and then gradually decreases even though the mental work may continue. These data are followed by a discussion of variations in tonus patterns, possible differences between tonus of effort and affect and that of postural adjustments, and the correlation of increased knee-jerk reflex with increased tonus during mental work.

375 **Goodenough, Florence L.** "The Expression of the Emotions in Infancy." *Child Development* 2 (1931):96-101. *P T*

College students judged photographs of a ten-month-old child in various expressions and selected which description of the emotion and which situation best fitted each picture. The photographs were matched correctly (i.e., relative to what the photographer reported) 47 percent of the time, with many of the errors actually close to the correct situations. The author argues against imitation or learning and for "native reaction patterns" of expression recognizable in infancy.

376 **Goodenough, Florence L.** "Expression of the Emotions in a Blind-Deaf Child." *Journal of Abnormal and Social Psychology* 27 (1932):328-333. *P*

A beautiful account of the expressive behavior of a ten-year-old girl, blind and deaf from birth, who had received little training. One incident in which a doll was dropped down her dress and she went through a series of expressions described as surprise, interest, disappointment, exasperation, rage, delight, and satisfaction is illustrated with photographs and presented as evidence for "native" or unlearned expression resembling that of normal children. Other characteristic expressions and movements are described, such as her unique way of dancing to entertain herself.

377 **Goodenough, Florence L., and Smart, Russell C.** "Inter-Relationships of Motor Abilities in Young Children." *Child Development* 6 (1935):141-153. *T*

A series of motor tasks were given to groups of children, ages two and one-half to five and one-half, and the results were factor-analyzed to explore the possibility of a "general factor of motor ability." The tasks included walking a path quickly and accurately, finger tapping, threading a needle, reaction time, and a "three-hole thrust" task of putting a stylus into three holes as quickly as possible. Tentative evidence of sex differences, improvement with age, and two factors, general motor maturity and attentiveness or carefulness, was found.

378 **Gostynski, E.** "A Clinical Contribution to the Analysis of Gestures." *International Journal of Psycho-Analysis* 32 (1951):310-318.

Probably the most detailed and elaborate case description available of gestures observed over several years in psychoanalytic sessions. Interpretation of the gestures, very exact description, and discussion of their origins in early childhood conflicts and their function as displacement and defense are presented. A woman patient described as having severe problems of penis envy is discussed, and her hand gestures and variations in the movement of right and left are minutely analyzed within a psychoanalytic framework.

379 **Gottschalk, Louis A.; Serota, Herman M.; and Shapiro, Louis B.** "Psychological Conflict and Neuromuscular Tension: I, Preliminary Report on a Method, as Applied to Rheumatoid Arthritis." *Psychosomatic Medicine* 12 (1950):315-319. *T*

EMG recordings of muscle tension in four groups (arthritic patients with and without analytic treatment, cardiovascular patients not in treatment, and controls who were in analysis) suggest that the level of muscle tension differs according to whether the person was in treatment, but not according to somatic illness.

380 **Graham, Frances K.** "Behavioral Differences between Normal and Traumatized Newborns: I, The Test Procedures." *Psychological Monographs*, Whole No. 427, 70 (1956):1-16. *T*

This paper reports on test procedures used in a neonate research project, four measures of which are relevant here because they include explicitly defined movement behaviors: the Visual Scale, the Maturational Scale, and the Irritability and Muscle Tension Ratings. Recording forms are reprinted here, and the Muscle Tension Rating is particularly detailed.

381 **Graham, Frances K.; Matarazzo, Ruth G.; and Caldwell, Bettye M.** "Behavioral Differences between Normals and Traumatized Newborns: II, Standardization, Reliability and Validity." *Psychological Monographs*, Whole No. 428, 70 (1956):1-33. *G T*

The tests described in the previous article were given to 265 normal and 81 traumatized infants, the latter diagnosed as having brain injury, and proved to discriminate well between the two groups and to assess the degree of trauma even in cases where the experimenters did not know the pediatrician's diagnosis. Age norms are presented. Test-retest and interscorer agreement was high; test intercorrelations were relatively low. Negro infants were superior to whites on the Maturational and Vision Scales, but sex and socioeconomic differences did not affect performance.

382 **Graham, James D. P.** "Static Tremor in Anxiety States." *Journal of Neurology, Neurosurgery and Psychiatry* 8 (1945):57-60. *G T*

Following a review of research on tremor, definitions of different types, neurophysiological theories concerning its causes, and means for measuring it, research is reported on the differences between static-tremor curves obtained from a normal group and from a group of men considered to have anxiety states. A measuring apparatus that optically magnified any finger tremor while the hand was at rest is described.

383 **Grant, E. C.** "An Analysis of the Social Behavior of the Male Laboratory Rat." *Behaviour* 21 (1963):260-281. *D*

An elaborate pattern analysis of agressive-submissive behaviors in male rats, tracing sequences of postures and acts and deriving diagrams of aggressive and flight pathways relative to specific postures. The territorial and social functions of the postures and approach patterns are discussed.

384 **Grant, E. C., and Mackintosh, J. H.** "A Comparison of the Social Postures of Some Common Laboratory Rodents." *Behaviour* 21 (1963):246-259. *D*

Analyzing the postures of rodents when one is put into the cage of another, the authors carefully describe and illustrate with drawings what the actual positions and actions are. They group them in terms of introductory acts, flight movements, aggressive elements, mating elements, ambivalence postures, and displacement activities and discuss their significance.

385 **Grant, Ewan C.** "An Ethological Description of Non-verbal Behavior during Interviews." *British Journal of Medical Psychology* 41 (1968):177-183. *D*

Three groups of subjects—chronic schizophrenic patients, neurotic patients, and students—were observed in interviews, and the experimenter recorded their speech simultaneously with his body movement observations in code words. Four groups of behaviors are discussed as elements of flight, assertive, relaxed, or contact. Flight behaviors and "restriction of repertoire" were more prominent in the seriously ill patients, although the author states that nonverbal behavior has the "same structure" in each group.

386 **Grant, Ewan C.** "Human Facial Expression." *Man* 4 (1969):525-536. *P*

Essentially a publication of a checklist for ethological study of behavior that includes more than facial expression. The 118 items in the list, organized according to body part involved, vary from definitions of blinking to crouching. Certain items are grouped under "aggressive" or "flight" elements, and some are illustrated with photographs of children.

387 **Greenacre, Phyllis.** "Infant Reactions to Restraint: Problems in the Fate of Infantile Aggression." *American Journal of Orthopsychiatry* 14 (1944):204-218.

The effects of physical and/or psychological restraint in childhood on personality development, particularly suppression of rage, are considered. Clinical examples, experimental literature, and culture customs of swaddling and head binding are assessed.

388 **Greene, Donovan R.** "The Effects of Aging on the Component Movements of Human Gait." Ph.D. dissertation, University of Wisconsin, 1959. 129 pp. (Datrix order no. 59-3192)

Gait—normal, accelerated, decelerated, and normal but without shoes—was measured in a group of 150 subjects ranging from four to seventy-one years of age. Analysis of measures of travel time and contact time indicated that normal pace remains fairly consistent from twelve to seventy; that there are sex differences; that there is little difference in an individual's gait from day to day; and that individual differences within the same age group generally are greater than differences between age groups.

389 **Grinstein, Alexander.** *The Index of Psychoanalytic Writings,* vol. 5, *Subject Index.* New York: International Universities Press, 1960. pp. 2223-2802.

More articles on psychoanalytic interpretation of body movement, motor development, and abnormal movement behavior in relation to psychopathology may be found in this index.

390 **Gross, Leonard R.** "Effects of Verbal and Nonverbal Reinforcement in the Rorschach." *Journal of Consulting Psychology* 23 (1959):66-68. *T*

Head nods or saying "good" proved to increase the number of human responses on the Rorschach.

391 **Grotowski, Jerzy.** *Towards a Poor Theatre.* Holstebro, Denmark: Odin Teatrets Forlag, 1968. 262 pp. *D P* [New York: Simon & Schuster (Clarion Books), 1969]

There are chapters on actor training that are a series of body movement and vocal excercises reflecting this noted theatrical director's attention to and understanding of body expression.

392 **Guilford, J. P.** "An Experiment in Learning to Read Facial Expression." *Journal of Abnormal and Social Psychology* 24 (1929):191-202. *G T*

Subjects given brief training in judgment of facial expression improved by 51 percent over their original scores and became "more uniform in ability." Differences between "analytical" and "unanalytical" judges and the effects of exposure time were also noted.

393   Guilford, J. P. "A System of the Psychomotor Abilities." *American Journal of Psychology* 71 (1958):164-174.

Reviewing literature on motor abilities, the author derived a two-way classification of psychomotor factors based on (1) the part of the body involved and (2) the type of ability: strength, impulsion, speed, static and dynamic precision, coordination, and flexibility.

394   Guilford, J. P., and Wilke, Margaret. "A New Model for the Demonstration of Facial Expressions." *American Journal of Psychology* 42 (1930):436-439. *D*

Description of a full-face demonstration model in which different expressions could be illustrated by piecing together different facial parts or features.

395   Haggard, Ernest A., and Isaacs, Kenneth S. "Micro-Momentary Facial Expressions as Indicators of Ego Mechanisms in Psychotherapy." In *Methods of Research in Psychotherapy*, edited by L.A. Gottschalk and A. H. Auerbach, pp. 154-165. New York: Appleton-Century-Crofts, 1966. *G T*

The authors report data on micromomentary facial expressions (MME's), those observed to occur and disappear within 1/5 to 1/8 of a second in their research of psychotherapy films. Problems of observer reliability are extensively discussed, along with tentative findings of individual differences in MME's and a relationship between an increase in MME's and active emotional conflict. This report is particularly valuable for the frankness and detail with which the investigators describe how they actually went about the research, what problems they encountered, how they assessed observer agreement, and how they conceptualized what they found.

396   Hall, Edward T., Jr. "The Anthropology of Manners." *Scientific American* 192 (Apr. 1955):84-90. *D*

An account of cross-cultural differences in manners—particularly punctuality and distances maintained between speakers—in South America, North America, and the Middle East, with a number of examples of manners involving body movements.

397   Hall, Edward T. "A System for the Notation of Proxemic Behavior." *American Anthropologist* 65 (1963):1003-1026. *N*

Hall summarizes the nature and origin of proxemic research (the study of space between people in groups and the spatial organization of buildings and towns) and presents a simple notation system useful for field studies. The system includes symbols for three "basic postures," sex identification, spatial orientation of individuals to each other, spatial distance, and distance in terms of degrees of touching, eye contact, heat and odor detection, and voice volume.

398   Hall, Edward T. *The Hidden Dimension.* New York: Doubleday & Company, 1966. 201 pp. *D P T*

One of Hall's presentations of proxemics and the sociological and communicational significance of spacing patterns in animals and man. Included is discussion of patterns and functions of distance regulation and the effects of crowding in animals; space and the distance and immediate sense receptors; analysis of distances in man; and proxemic patterns of Germans, English, French, Arabs, and Japanese.

399 Hall, Edward T. "Proxemics." *Current Anthropology* 9 (1968):83-108. *D P T*

A review of the historical background, conceptual framework, and descriptive system of proxemics, with notes on recent research and the author's work in progress. Also included are critical comments by seventeen anthropologists and linguists.

400 Hall, K. R. L., and De Vore, Irven. "Baboon Social Behavior." In *Primate Behavior: Field Studies of Monkeys and Apes,* edited by I. De Vore, pp. 53-110. New York: Holt, Rinehart and Winston, 1965. *G P T*

Dominance and mating behaviors are analyzed in detail. There is a table of facial and body expressions and tactile behavior: when they occur, who does them, and what their significance is, together with descriptions of behaviors in response to threat, attack, surprise, and friendly encounters and greetings.

401 Halverson, H. M. "An Experimental Study of Prehension in Infants by Means of Systematic Cinema Records." *Genetic Psychology Monographs* 10 (1931): 107-286. *D G P T*

Frame-by-frame film analysis of groups of twelve or more infants at sixteen, twenty, twenty-four, twenty-eight, thirty-two, thirty-six, forty, and fifty-two weeks of age who are filmed from the waist up as they manipulate a cube. The paths of the arms; patterns of visual regard; types of approach in terms of time, height, body parts, and aim (backhand, circuitous, straight, etc.); laterality; types of rotation, grasping, and lifting; and the things done with the cube are all carefully described and illustrated, and the age differences are discussed in this exhaustive study.

402 Halverson, H. M. "Studies of the Grasping Responses of Early Infancy: I." *Journal of Genetic Psychology* 51 (1937):371-392. *T*

Discussion of the nature and development of the grasping reflex and voluntary prehension in infants; the period of maximum reflex strength; and experiments on the infant's grasping a smooth or rough rod and reaction to light pressure.

403 Halverson, H. M. "Studies of the Grasping Responses of Early Infancy: II." *Journal of Genetic Psychology* 51 (1937):393-424. *G P T*

A report on age, sex, and handedness differences in the clinging strength of infants from birth to fifty-two weeks, as well as patterns of pressing and degree of hand and finger pressure at different ages. Testing apparatus is described.

404 Halverson, H. M. "Studies of the Grasping Responses of Early Infancy: III." *Journal of Genetic Psychology* 51 (1937):425-449. *G P T*

The experiments show that variations in the grasp response are a function of the amount of activity and the posture and arm position of the infant at the time the hand is stimulated, as well as of the size of the hand and the diameter of the rod grasped. At the end of this article, the author summarizes the results of his research on the grasping reflex.

405 Halverson, H. M. "Infant Sucking and Tensional Behavior." *Journal of Genetic Psychology* 53 (1938):365-430. *D G T*

Patterns of sucking pressure of infants were measured with an apparatus inserted into the rubber nipple; grip pressure was also recorded on a kymograph, and observations were made of the baby's general activity, muscle tension, changes in posture, etc. Reactions to "difficult" and "easy" nipples were assessed. Sucking rate, duration, and type and degree of pressure are described; kymographs of the sucking and gripping patterns are presented, and body movements during feeding and frustration of feeding are described. The circumstances in which penile tumescence occurred and declined are extensively analyzed. The article concludes with a discussion of the experimental results, the nature of bodily tension, and the patterns of penile tumescence.

406 **Halsman, Philippe.** *Jump Book.* New York: Simon and Schuster, 1959. 94 pp. *P*

Photographs of many famous people jumping. However tongue-in-cheek, the noted photographer's impressive collection and his comments on the relation between jumping and personality are intriguing. Judging from the rich variation of jumping styles, perhaps "jumpology" should be taken seriously.

407 **Hamalian, Leo.** "Communication by Gesture in the Middle East." *ETC: A Review of General Semantics* 22 (1965):43-49. *D*

The author, a semanticist, describes specific gestures he observed among Arabs in Syria, Lebanon, and Jordan and what he found them to mean.

408 **Hambly, W. D.** *Tribal Dancing and Social Development.* New York: The Macmillan Company, 1927. 296 pp. *D P*

Contains an extensive, if dated, survey and bibliography of dance ethnology; description of dances and ceremonies of "primitive" cultures around the world; and theories about the origins and functions of dance in society. Included are chapters on music, musical instruments, dance in ancient history, and dances of war, initiation, courtship, harvest, magic, and death.

409 **Hanawalt, Nelson G.** "The Role of the Upper and Lower Parts of the Face as a Basis for Judging Facial Expressions: I, In Painting and Sculpture." *Journal of General Psychology* 27 (1942):331-346. *T*

Subjects of this experiment checked which of six emotion categories best described the painting or sculpture shown to them. Those who were judging full faces were in fair agreement; judgments of only the upper face were about as good as those for the lower face; and sculpture was more difficult to judge than painting was.

410 **Hanawalt, Nelson G.** "The Role of the Upper and the Lower Parts of the Face as a Basis for Judging Facial Expressions: II, In Posed Expressions and 'Candid Camera' Pictures." *Journal of General Psychology* 31 (1944):23-36. *T*

Subjects selected which of six emotion categories best described the photograph. In general, the lower half of the face was as good as the upper in yielding agreement; but for particular types, the lower was better for happy expressions, and the upper for surprise and fear. Most confusion was found between recognition of expressions of happiness and pain in the upper half.

411 **Hanna, Judith L.** "African Dance as Education." In *Impulse: Annual of Contemporary Dance*, pp. 48-56, 1965. *P*

A paper on how various African dances transmit the culture and educate the children in role differentiation, values, and the society's knowledge.

412 **Hardyck, Curtis D.** "Personality Characteristics and Motor Activity: Some Empirical Evidence." *Journal of Personality and Social Psychology* 4 (1966): 181-188. *T*

EMG recordings were made from forearm and forehead while twenty subjects took either the Rorschach Test or the Guilford Consequences Test. Subjects were given the California Psychological Inventory to explore relationships between certain personality characteristics and muscle tension. No relation was found between the degree of M responses on the Rorschach and muscle tension, or between high scores on the self-acceptance, socialization, self-control categories, etc., and low muscle tension.

413   **Hare, A. Paul, and Bales, Robert F.** "Seating Position and Small Group Interaction." *Sociometry* 26 (1963):480-486. *T*

Studying the seating arrangements and behavior of "five-man laboratory groups" in task and social sessions, the authors describe a number of relationships between position and dominance, interaction patterns, and personality characteristics.

414   **Harris, C. Stanley; Thackray, Richard I.; and Schoenberger, Richard W.** "Blink Rate as a Function of Induced Muscular Tension and Manifest Anxiety." *Perceptual and Motor Skills* 22 (1966):155-160. *T*

Eye blinks and physiological activity were measured in twenty-five young men during relaxation and during muscle tension elicited by gripping a dynomometer. Results indicate that the blink rate does reflect generalized muscle tension and does correlate with measures of anxiety such as the Taylor Manifest Anxiety Scale.

415   **Harris, Marvin.** *The Nature of Cultural Things.* New York: Random House, 1964. 209 pp. *G T*

Particularly relevant for the definition and method of classifying "actones," considered the smallest behavioral bits and consisting of body motion and its effect on the environment (pp. 36-52). A schema is presented for developing an "actonemic vocabulary" listing the actoneme, body part, motion, and environmental effect. Methods of describing behavior in sequence are presented.

416   **Harrison, Jane E.** *Ancient Art and Ritual.* New York: Henry Holt and Company, 1913. 256 pp. [Westport, Conn.: Greenwood Press, 1951]

A treatise on the relationship between art and ritual, using ancient Greek drama as a focus.

417   **Harrison, Randall P.** "Pictic Analysis: Toward a Vocabulary and Syntax for the Pictorial Code, with Research on Facial Communication." Ph.D. dissertation, Michigan State University, 1964. (Datrix order no. 65-6079)

Subjects were given schematic drawings of faces that varied in the position of eyebrows, eyes, and mouth; they judged them according to which terms for various moods, interpersonal relationships, demographic characteristics, etc., were appropriate. Harrison found that specific variations were sufficient to get certain responses (e.g., downturned eyebrow = angry).

418   **Harrison, Randall P.** *Nonverbal Communication.* Englewood Cliffs, N.J.: Prentice-Hall, forthcoming.

419   **Harrison, Randall P.** "Nonverbal Communication." In *Communications Handbook*, edited by I. Pool, W. Schramm, N. Maccoby, F. Fry, and E. Parker. Chicago: Rand McNally & Company, forthcoming.

420   **Harrison, Randall P.** "The Nonverbal Approach." In *Approaches to Human Communication*, by R. Budd and B. Ruben. New York: Spartan Books, forthcoming.

421   **Harrison, Wade; Lecrone, Harold; Temerlin, Maurice K.; and Trousdale, William W.** "The Effect of Music and Exercise upon the Self-Help Skills of Nonverbal Retardates." *American Journal of Mental Deficiency* 71 (1966):279-282. *T*

Severely retarded children given exercises with music for twenty minutes a day for four weeks did significantly better on self-help skills than did those given only exercise, or music without exercise, or neither.

422 Hastorf, Albert H.; Osgood, Charles E.; and Ono, Hiroshi. "The Semantics of Facial Expression and the Prediction of the Meanings of Stereoscopically Fused Facial Expressions." *Scandinavian Journal of Psychology* 7 (1966):179-188.

Photographs of facial expressions were judged on a number of terms for each of three scales: pleasantness, activation, and control. A factor analysis showed high loadings for pleasantness and activation terms, but "hard-soft," "tight-loose," and "open-closed" did not prove to define the control dimension. A different factor was suggested. The judgments of subjects who were shown two different photographs stereoscopically were also analyzed.

423 Hayes, Francis. "Gestures: A Working Bibliography." *Southern Folklore Quarterly* 21 (1957):218-317.

Most of the references included in the Hayes bibliography are not duplicated in the present bibliography. Mostly annotated, note is also made of where some of the sources may be found. In contrast to this body movement bibliography, Hayes cites many foreign-language sources, rare historical documents, and literary references as well as comments on gestures from the popular press. He also cites numerous references to sign languages, ritual movements, and such special types of gestures as the gesture signals used in television.

424 Hazzard, Florence W. "Development of an Infant in Grasping, Reaching and Body Movements." *Child Development* 2 (1931):158-160.

A brief article listing prehensile behaviors observed in an infant from birth to 176 days of age.

425 Heath, S. Roy. "Rail-Walking Performance as Related to Mental Age and Etiological Type among the Mentally Retarded." *American Journal of Psychology* 55 (1942):240-247. *G T*

Difficulty in locomotor balance clearly distinguished endogenously retarded boys from those considered exogenous.

426 Hebb, D. O. "The Forms and Conditions of Chimpanzee Anger." *Canadian Psychological Association Bulletin* 5 (1945):32-34.

Arguing that anger is recognized from "its relation to a temporal sequence of behavior," and not from facial expression, Hebb describes situations in which chimpanzees reacted with rage that are as subtle and complex as would be true for humans.

427 Hebb, D. O. "Emotion in Man and Animals: An Analysis of the Intuitive Processes of Recognition." *Psychological Review* 53 (1946):88-106.

A discussion of the analysis and recognition of emotions in primates and humans. Arguing that recognition of emotions in man or ape is based on observation of deviations in overt behavior once the observer knows the "habitual base line," and using descriptions and interpretations of chimpanzee expressions and actions in the Yerkes labs, the author presents a critique of prior research on recognition of expression and an analysis of definitions of emotion.

428 Hediger, H. *Studies of the Psychology and Behaviour of Captive Animals in Zoos and Circuses.* London: Butterworth Scientific Publications, 1955. 166 pp. *P T*

A great deal of information about the behavior and body expression of very diverse animals. There are intriguing descriptions of the everyday life of the animals in captivity; their territorial behavior; interaction between males, males and females, and mothers and infants; animal-human relationships; and the role of signals and distance in animal training. For a current source see: H. Hediger, *Psychology and Behavior of Animals in Zoos and Circuses,* translated by G. Sircom. New York: Dover Publications, 1969.

429 **Henley, Nancy M.** "The Politics of Touch." Presented at a meeting of the American Psychological Association, n.p., 1970. 7 pp. (Copies from author, c/o Department of Psychology, University of Maryland—Baltimore County, 5401 Wilkens Avenue, Baltimore, Md. 21228)

In a paper presented as part of a discussion on "Social Psychology and Women's Liberation," the author describes different situations and nonverbal ways in which a man asserts his status over a woman. From observations in a public place, data is presented showing that men initiate touch far more than women, that older persons touch younger ones more, and that those of a higher socioeconomic status touch those of a lower status, but rarely vice versa. The author discusses this as support for her thesis that man's greater freedom of initiating touch reflects his superior status over women.

430 **Hess, Eckhard H.** " 'Imprinting' in Animals." *Scientific American* 198 (Mar., 1958):81-90. *D G P*

A discussion of "imprinting," in which for certain newborn animals the first moving object they see elicits immediate and later social behaviors. Experiments on imprinting in ducks, to different objects and at critical periods, are reported.

431 **Hess, Eckhard H.** "Pupillometric Assessment." In *Research in Psychotherapy,* vol. III, edited by J. M. Schlien, pp. 573-583. Washington, D.C.: American Psychological Association, 1968. *T*

"Pupillometric assessment" is the study of positive and negative attitudes and affects through the degree of eye-pupil dilation. Hess begins with a review of research, particularly on abnormalities of pupil dilation, and psychopathology and includes a valuable bibliography on the subject. He then summarizes most of his and his colleagues research on pupillometric assessment. An apparatus for observing dilation size while subjects look at pictures is described. Among the results are studies showing that heterosexual males dilate to photographs of females, homosexual males to male photographs, and that women's pupils dilate more to pictures of men; heterosexual males dilate more to pictures of women who themselves have larger pupils; subjects' pupils dilate more to pictures of food they prefer; there are correlations between patterns of pupil dilation and GSR patterns.

432 **Hewes, Gordon W.** "World Distribution of Certain Postural Habits." *American Anthropologist* 57 (1955):231-244. *D*

From extensive study of photographs and relevant literature, the author describes and illustrates 100 of the most common patterns of sitting, kneeling, squatting, and standing; also, which sex performs them and in which parts of the world they predominate. Factors influencing these positions are discussed.

433 **Hewes, Gordon W.** "The Anthropology of Posture." *Scientific American* 196 (Feb., 1957):123-132. *D*

Based on study of photographs and ancient art, maps of the distribution of "static postures" characteristic of men and women and drawings of typical sitting and standing postures in different cultures around the world (including some ancient ones) are presented, with a discussion of their features and origins.

434   Hinde, Robert A. *Animal Behaviour: A Synthesis of Ethology and Comparative Psychology.* New York: McGraw-Hill Book Company, 1966. 534 pp. *D G T*

An ambitious and valuable synthesis of animal research, which focuses on immediate causation of behavior and developmental and learning processes in vertebrates. Of particular note here are the chapters on classification of behavior, control of movement, spontaneity and rhythmicity, motivational mechanisms, courtship and threat behavior, and "Development: The Form and Orientation of Movement Patterns." The book is extensively illustrated and has a sixty-page bibliography.

435   Hinde, R. A., and Rowell, T. E. "Communication by Postures and Facial Expressions in the Rhesus Monkey (Macaca Mulatta)." *Proceedings of the Zoological Society of London* 138 (1962):1-21. *P*

Observations of zoo monkeys, including their sitting postures, aggressive and threat movements, fear responses, and friendly behavior, and the contexts in which these occur, are described and illustrated.

436   Hinde, R. A., and Tinbergen, N. "The Comparative Study of Species-Specific Behavior." In *Behavior and Evolution,* edited by A. Roe and G. G. Simpson, pp. 251-268. New Haven, Conn.: Yale University Press, 1958.

An analysis of the evolution and function of display movements in various animal species.

437   Hindley, C. B. "The Griffiths Scale of Infant Development: Scores and Predictions from 3 to 18 Months." *Journal of Child Psychology and Psychiatry* 1 (1960):99-112. *T*

The Griffiths Scale includes motor tasks and rating procedures for locomotion, eye and hand movements, and "performance" of infants, as well as for hearing and speech and personal-social behavior. In the study reported here, in which the rating applied to a group of English infants, scores were too unstable to have predictive value, although the author contends that the scale may be useful in studying same-age infants.

438   Hoffman, Stanley P. "An Empirical Study of Representational Hand Movements." Ph.D. dissertation, New York University, 1968. 151 pp. *G P T* (Datrix order no. 69-7960)

A study of "literal" hand movements that "depict some aspect of the manifest content of speech" (p. 3), called here "representational hand movements" (RHM's). Twenty-four young women participated in a study in which they recalled stories and described TAT pictures and Rorschach cards from memory in a face-to-face position, then back-to-back, and finally face-to-face with hands restrained. Results showed that the rate of hand movements related to verbal fluency; that RHM's may be related to attempts to express diffuse visual imagery (there were more with the Rorschach than with the TAT); and that inhibition of RHM's slightly affects speech.

439   Holbrook, Sarah Fitch. "A Study of the Development of Motor Abilities between the Ages of Four and Twelve, Using a Modification of the Oseretsky Scale." Ph.D. dissertation, University of Minnesota, 1953. 195 pp. (Datrix order no. 5537)

The author established tentative norms for motor development level through administration of a modified version of the Oseretsky Tests of Motor Proficiency to groups of forty children at each age (4-12) matched by sex and socioeconomic status. With increasing age, there was an increase in the percentage of those passing, as expected; but the author asserts the need for good validation and factor studies. She did find individual motor consistency over two years; no correlation between motor performance and sex, socioeconomic level, or I.Q.; and a tendency for the "well-adjusted" subjects to perform better.

440 **Honkavaara, Sylvia.** "The Psychology of Expression." *British Journal of Psychology: Monograph Supplements* 32 (1961):1-96. *D G P T*

A formidable study of perception of emotion formulated within a dynamic. developmental viewpoint similar to Heinz Werner's. It has an excellent bibliography and review of literature on theories of emotion and recognition of expression. Experiments are reported that support the hypothesis that there are four stages of development in perceptual terms of the following order: dynamic-affective (subject and object are not distinguished); matter-of-fact (able to perceive objects and actions); physiognomic (able to perceive emotion and physiognomic qualities); and intersensory property (able to perceive style, to understand people, to be "intuitive"). A number of experiments using Finnish or English subjects as young as three years old and as old as eighty were conducted. In one, for example, children and adults from ages five to forty-five were asked to describe pictures. The "expression" answers increased with age, whereas the "matter-of-fact" answers decreased correspondingly, indicating to the author that recognition of emotion is not "instinctive" or socially learned but is primarily the result of mental maturation, and also that agreement in recognition is a function of mental development. In addition to experiments with recognition of emotion from facial and body expression using photographs or drawings, there are studies of the physiognomic character of words, drawings, sounds, writing style, and so on.

441 **Horst, Louis.** *Modern Dance Forms in Relation to the Other Modern Arts.* (Impulse Publications, 1961; reprinted., Brooklyn, N.Y.: Dance Horizons, n.d.). 149 pp. *D P*

An excellent treatise on modern dance choreography and aesthetics by a prominent musician and teacher of choreography. In discussing how the modern dancer should draw on historical, ethnic, and artistic sources, the author says a great deal about movement styles, from those reflected in medieval art to those observed in rural America.

442 **Howe, Clifford E.** "A Comparison of Motor Skills of Mentally Retarded and Normal Children." *Exceptional Children* 25 (1959):352-354. *T*

Normal and retarded schoolchildren were given a battery of motor tests (involving jumping, balance, tracing, grip strength, running, etc.). The normal children did consistently better. Of note was the inability of the retarded children to balance on one foot, although they did not show organic damage in neurological tests.

443 **Huber, Ernst.** *Evolution of Facial Musculature and Facial Expression.* Baltimore: The Johns Hopkins Press, 1931. 184 pp. *D*

This excellent, well-illustrated work begins with a history of research in facial musculature and expression. Analysis of the facial musculature in lower vertebrates and mammals is followed by extensive assessment of its evolution in primates and man. Racial and cultural differences in facial anatomy and expression are assessed, and developmental aspects are noted.

444   Hulin, Wilbur S., and Katz, Daniel. "The Frois-Wittmann Pictures of Facial Expression." *Journal of Experimental Psychology* 18 (1935):482-498. *P T*

The seventy-two Frois-Wittmann photographs are reprinted here along with a study in which subjects sorted the pictures into groups of similar expressions. There was wide scatter and some agreement.

445   Humphrey, Doris. *The Art of Making Dances.* New York: Rinehart and Company, 1959. 189 pp. *D P* [New York: Grove Press (Evergreen Books), 1962]

A readable book with many examples and illustrations of actual body movements. Particularly relevant to a study of movement behavior are the following sections: movement dynamics in relation to mood and an individual dancer's temperament; rhythm in relation to weight, breathing, and emotion; expression of emotion in dance and how this choreographer interprets it; gestures (social, functional, ritual, and emotional) and how variations of a gesture suggest different motivations and feelings.

446   Hunt, Valerie. "The Biological Organization of Man to Move." In *Impulse: The Annual of Contemporary Dance*, pp. 51-62, 1968.

The author discusses neurophysiological, reflexive, and psychological aspects of body movement, from data based on electromyographic research of muscle tension patterns. Her findings about various pattern and coordination aspects of muscle tension and force, space, and time variations in movement are analyzed in relation to individual movement styles, body image, motor performance, and emotion.

447   Hunt, Valerie. "Neuromuscular Structuring of Human Energy." Presented at the Forty-fifth Conference Program and in the Annual Report of the Western Society for Physical Education of College Women, n.p., 1970. 8 pp.

A glimpse at some revolutionary developments in electromyography and the study of patterning in neuromuscular activity. The author describes research at the Movement Behavior Laboratory at U.C.L.A. Miniaturized, telemetered EMG instruments adapted from aerospace science are used there which can record everything from activity of large muscle groups to lips while the individual moves without restriction. These instruments and new techniques in placement are discussed. Thus far, four patterns of energy release (undulating, sustaining, burst, and restrained) have been studied, with resulting evidence that they differ by situation and by cultural group. Advanced computer analysis is being made of the pattern frequencies, with indications that it may be possible to determine which movements arise from the spinal level and which from the brainstem and cerebral cortex! Possible applications of such research are discussed in this report.

448   Hunt, William A. "Studies of the Startle Pattern: II, Bodily Pattern." *Journal of Psychology* 2 (1936):207-213. *P T*

Twenty-nine subjects were photographed along with equipment designed to objectively measure their movement during a gunshot. Typical and individual characteristics of head and body movement are described, along with analysis of the effects of repetition.

449   Hunt, William A.; Clarke, Frances M.; and Hunt, Edna B. "Studies of the Startle Pattern: IV, Infants." *Journal of Psychology* 2 (1936):339-352. *P T*

Sixty infants from eight days to eighteen months old were filmed during response to a blank gunshot. From film analysis, the forms and variations of the Moro reflex and the startle pattern are reported in terms of how different body parts are moved. Possible relationships between the Moro response, which dissipates at about five months of age, and the later startle patterns are discussed.

450 Hunt, W. A.; Landis, C.; and Jacobsen, C. F. "Studies of the Startle Pattern: V, Apes and Monkeys." *Journal of Psychology* 3 (1937):339-343. *P T*

The reactions to gunshots of nine chimpanzees and six monkeys were filmed and were found to be similar to the startle pattern in humans. In primates the pattern is more intense, involves more body parts, suggests a protective action, and in some ways is more similar to that of human infants than of human adults.

451 Huntley, C. W. "Judgments of Self Based Upon Records of Expressive Behavior." *Journal of Abnormal and Social Psychology* 35 (1940):398-427. *G T*

Fascinating experiments in which subjects' judgments of pictures of their own hands, profiles, tapes of their voices, etc., were more favorable than judgments of or by others, although the pictures and tapes were disguised so that they did not recognize them as their own. Unfortunately, it does not include self-judgments of one's own movement.

452 Huston, Paul E., and Shakow, David. "Studies of Motor Function in Schizophrenia: III, Steadiness." *Journal of General Psychology* 34 (1946):119-126. *T*

Male schizophrenic patients and normal controls performed a stylus test as a measure of motor steadiness and fine coordination. The patients in general performed more poorly, but those patients who cooperated well did as well as the normal group.

453 Hutchinson, Ann. *Labanotation or Kinetography Laban: The System of Analyzing and Recording Movement.* Rev. and expanded ed., New York: Theatre Arts Books, 1970. 528 pp. *D N*

An extremely comprehensive presentation of Labanotation, or Kinetography Laban, first invented by Rudolf Laban and currently the most widely used dance notation. Although this is primarily a textbook for dancers and dance notators, attention is given to the system's potential in recording any movement. The book is replete with examples of movement notation, a glossary of symbols, and illustrations of the concepts involved in the system.

454 Hutt, Corinne, and Ounsted, Christopher. "The Biological Significance of Gaze-Aversion, with Particular Reference to the Syndrome of Infantile Autism." *Behavorial Science* 11 (1966):346-356. *D G*

Systematic observations of eight autistic children, ages three to six years, showed that although other components of their social behavior might appear normal (such as reaching to be picked up), the children averted their gaze, lowered their head, and avoided eye contact continuously. In a room that had drawings mounted on 3-foot stands (happy, sad, and blank human faces; monkey and dog faces), the autistic children spent more time encountering the picture-stand of the dog than of the humans, whereas nonautistic children fixated more on the happy human face. Diagrams of the exploratory patterns of the children are presented. On the basis of this study and another, in which autistic children showed almost no aggressive behavior and yet were never attacked in a room crowded with active children, the authors discuss the function of gaze aversion in minimizing arousal and appeasing others.

455 *Impulse 1969-1970: Extensions of Dance.* San Francisco: Impulse Publications, 1970. 97 pp. *D P*

This edition of *Impulse* has a number of articles on the psychology and anthropology of body movement: notably, "The Dancing Healers of Ceylon" by John Halverson; "An Anthropologist Looks at Ballet as a Form of Ethnic Dance" by Joann Kealiinohomoku; brief discussions of Japanese, Indian, and African dance; and articles on the therapeutic use of body movement by Mary Whitehouse, Gay Cheney, Rhoda Winter Russell, Varda Razy, and Joseph Schlicter.

456 **Irwin, Francis W.** "Thresholds for the Perception of Difference in Facial Expression and Its Elements." *American Journal of Psychology* 44 (1932):1-17. *D G T*

Subjects judged diagrams (a circle with slightly curved or straight lines for "eyes" and "mouth") as curved up, down, or straight; or as a smile, a frown, or neutral. Precision and improvement of perception was better for judgments of curvature than of expression. Other aspects of the subjects' perception are reported.

457 **Irwin, Orvis C.** "The Amount and Nature of Activities of Newborn Infants under Constant External Stimulating Conditions during the First Ten Days of Life." *Genetic Psychology Monographs* 8 (1930):1-92. *G T*

The activity patterns of four neonates were continuously studied over the first ten days of life and recorded with a stabilimeter-oscillograph apparatus and by systematic observation of sounds, postures, general body movement, and head, facial, and limb activity. Definitions of movements and actions and the symbols used to record them are presented. The frequency of specific body part movements, in percentage and per hour on different days and for each child, is presented and shows an increase in movement of the anterior region. Oscillations of mass activity over time are assessed, also showing clear increases. The developmental and neurophysiological implications of the results are presented.

458 **Irwin, Orvis C.** "The Amount of Motility of Seventy-Three Newborn Infants." *Journal of Comparative Psychology* 14 (1932):415-428. *G P T*

The amount of activity of seventy-three newborns was measured with a stabilimeter connected to a polygraph recorder for a period of three and one-quarter hours on each of the first sixteen days of life. Results show a significant increase in activity on the fourth or fifth day; striking individual differences between infants; and six times as much motility during wakefulness as during sleep.

459 **Irwin, Orvis C.** "The Distribution of the Amount of Motility in Young Infants between Two Nursing Periods." *Journal of Comparative Psychology* 14 (1932):429-445. *G T*

The amount of activity of seventy-three infants (measured with stabilimeter and polygraph) in a three-hour period between feedings significantly increased until the next feeding, which is discussed here in relation to hunger.

460 **Jablonko, Allison Peters.** "Dance and Daily Activities among the Maring People of New Guinea: A Cinematographic Analysis of Body Movement Style." Ph.D. dissertation, Columbia University, 1968. 337 pp. *D P T* (Datrix order no. 69-3077)

A detailed and exhaustive study of the movement characteristics of the Maring of New Guinea and ways in which their dancing parallels their movements in daily life. A great deal of care has gone into the definition of units observed, ways of graphically recording the movement patterns, and frame-by-frame film analysis of various activities.

461 **Jackson, C. V.** "The Influence of Previous Movement and Posture on Subsequent Posture." *Quarterly Journal of Experimental Psychology* 6 (1954):72-78. *T*

An example of experimental study of kinesthetic perception and motor control without vision. In this case, subjects showed a consistent upward tendency in attempts to extend their arms horizontally and a tendency to overshoot, depending on previous movements.

462 **Jacobson, Edmund.** *Progressive Relaxation: A Physiological and Clinical Investigation of Muscular States and Their Significance in Psychology and Medical Practice.* 2d rev. ed. Chicago: University of Chicago Press, 1938. 494 pp. *D G P T*

A classic study of muscular tension in relation to thinking, physical and psychiatric disorders, and emotion. Progressive relaxation techniques for treating hypertension by producing extreme relaxation through development of body awareness and the "muscle sense" are described and illustrated. A history of the method and its effects on pain reactions, reflexes, mental activities, and thinking are presented, along with a review of theories of emotion and the relation of relaxation to emotion. Extensive analysis of the chemical and neurophysiological bases of muscular contraction and relaxation is concluded with principles about the relationship between relaxation and hypnosis or suggestibility. There are chapters on electrical measurement of contraction and relaxation in relation to mental activity and hypertension disorders, together with assessment of relaxation treatments.

463 **Jacobson, Edmund.** *Biology of Emotions: New Understanding Derived from Biological Multidisciplinary Investigation; First Electrophysiological Measurements.* Springfield, Ill.: Charles C Thomas, Publisher, 1967. 211 pp. *D G T*

An impressive sysnthesis of thirty-five years of research on emotion and neuromuscular patterns. The relation of muscle activity and tension to adaptation, thinking, and perception (the "perception-evaluation-tension" adaptive pattern) and to neurophysiological processes is discussed. Theories of emotion are critically evaluated. The author then supports his theory that all emotional and mental activity is accompanied by specific neuromuscular patterns with results of EEG and EMG studies, and he proposes new principles about the nature of emotion. An outcome study of twenty-five severely anxious patients, showing the beneficial effects of progressive relaxation treatments, is presented along with a chapter on evaluation and treatment of tension and anxiety in military personnel.

464 **Jaeger, Martha.** "Some Aspects of Relationship between Motor Coordination and Personality in a Group of College Women." Ph.D. dissertation, Columbia University, 1957. 99 pp. (Datrix order no. 21117)

In a study of the relationship between motor coordination and temperament, a number of tests were administered to a group of college women, including the Brace Motor Test, the MacQuarrie Mechanical Aptitude Test, the Bernreuter Personality Inventory, health tests, and others. Statistical analyses of data from the total group show no significant trends, but a study of the extremes on several tests shows a pattern between motor coordination, height and weight, and specific temperament factors that corroborates Sheldon's trichotomies of temperament and body type. Unique individual patterns are also noted.

465   Jarden, Ellen, and Fernberger, Samuel W. "The Effect of Suggestion on the Judgment of Facial Expression of Emotion." *American Journal of Psychology* 37 (1926):565-570. *G T*

Six Piderit models of facial expression (angry, dismayed, horrified, disdainful, disgusted, and bewildered) were shown to students, who were given the name of the emotion or had it elaborately demonstrated and explained through mime. They then rated each model for how well it expressed the intended emotion. Suggestion greatly increased agreement in all except judgment of the "angry" model.

466   Jay, Leticia. "A Stick-Man Notation." *Dance Observer* (Jan., 1957):7-8. *N*

A popularized article on Jay notation and the need for notation in dance. The original work on this system is rare and unpublished, but the reader can at least get an idea of the system from this article.

467   Jay, Phyllis C. *Primates: Studies in Adaptation and Variability.* New York: Holt, Rinehart and Winston, 1968. 529 pp. *D G P T*

An anthology of field studies of primates, replete with information about the communicative and expressive behavior of monkeys and apes. The following chapters are notable for their descriptions of movement behavior. Chapter 2, by K. R. L. Hall, has observations of the locomotion, postures, aggressive and sexual behaviors, greetings, mother-infant interaction, anxious behaviors, and the threat-attack "repertoire" of the patas monkeys. Chapter 6, by John O. Ellefson, is a log of the moment-to-moment behaviors of gibbon monkeys involved in a territorial conflict. Chapter 12, by Jane Van Lawick-Goodall, concerns the expressive movements of Gombe Stream chimpanzees and is reviewed by itself in this bibliography. Chapter 15, by Peter Marler, has an analysis of communication signals and movement behaviors which serve to effect and maintain spacing among primates. Chapter 17, by S. L. Washburn and D. A. Hamburg, analyzes the origins and functions of aggressive behavior in monkeys and apes. This book has two excellent bibliographies on ethology and primate behavior.

468   Jelliffe, Smith E. "The Parkinsonian Body Posture: Some Considerations in Unconscious Hostility." *Psychoanalytic Review* 27 (1940):467-479.

The author proposes that individuals with more severe forms of parkinsonism, such as the parkinsonian posture, are likely to have weak superegos and greater psychopathology. He presents a case study indicating that unconscious hostility and sadism may contribute to the motor disturbance.

469   Jenness, Arthur. "The Recognition of Facial Expressions of Emotion." *Psychological Bulletin* 29 (1932):324-350.

Starting with physiognomic studies of the late 1800's, a good review of the early literature on judgment of emotion from facial expression. Work of European researchers which is not available in English but which is germane to this area is reviewed, notably that of Piderit, Duchenne, Dumas, Rudolph, and Wundt. The issues regarding innate patterns of facial expression considering infant development and innate versus learned recognition are discussed. The author goes on to review literature dealing with specific topics such as sex differences in recognition and recognition ability and "social intelligence."

470   Jersild, Arthur. "Modes of Emphasis in Public Speaking." *Journal of Applied Psychology* 12 (1928):611-620. *T*

Subjects' recall of statements made by speakers who varied their emphasis by repetition, position of the phrase, type of phrase, loudness, slowing speech, pausing, raising the arm, or banging the table was used to assess effectiveness of the mode of emphasis. In the list of those modes which were effective, the movements were ranked lowest.

471 **Jerstad, Luther G.** *Mani-Rimdu: Sherpa Dance-Drama.* Seattle: University of Washington Press, 1969. 192 pp. *D P*

An analysis of the dance drama performed in Buddhist monasteries of the Sherpas of Nepal. The Sherpas' religion and philosophy and the origin, history, physical setting, and instruments of the Mani-Rimdu are described, followed by detailed description of its performance and significance. This book has a good bibliography on the religion, rituals, and dance drama of Nepal, China, and Tibet.

472 **Joffe, J. M.** *Prenatal Determinants of Behaviour.* Oxford, Eng.: Pergamon Press, 1969. 366 pp. *G T*

An impressive review of literature on the subject of prenatal determinants, including studies of the influences of maternal emotions on fetal activity.

473 **Johannesson, Alexander.** "Gesture Origin of Indo-European Languages." *Nature* 153 (1944):171-172.

Presentation of linguistic analyses to support Richard Paget's theory that language originated from imitation by the speech organs of hand gestures and body movements.

474 **Johnson, Buford J.** *Experimental Study of Motor Abilities of Children in the Primary Grades.* Baltimore: The Johns Hopkins Press, 1917. 63 pp. *G T*

Apparatus for tests of steadiness, tapping rate, precision, and reaction time are described. Practice effects, handedness, and sex differences are assessed.

475 **Johnson, Buford J.** "Changes in Muscular Tension in Coordinated Hand Movements." *Journal of Experimental Psychology* 11 (1928):329-341. *G T*

Muscle tension and pressure patterns of adults and children were studied with a tapping board and stylus apparatus. Differences and types of patterns are analyzed.

476 **Johnson, Granville B.** "A Study of the Relationship That Exists between Physical Skills as Measured and the General Intelligence of College Students." *Research Quarterly of the American Association for Health and Physical Education* 13 (1942):57-59.

No significant relationship was found between intelligence as measured and performance on the Johnson Physical Skill Test.

477 **Jones, Frank P.** "Method for Changing Stereotyped Response Patterns by the Inhibition of Certain Postural Sets." *Psychological Review* 72 (1965):196-214. *D P T*

Introspective reports and experimental data derived from multiple-image photography, X-ray photography, and EMG recordings of subjects moving from sitting to standing positions are presented to show the effects of guiding the head in such a way as to correct habitual postures.

478 **Jones, Frank P., and Hanson, John A.** "Time-Space Pattern in a Gross Body Movement." *Perceptual and Motor Skills* 12 (1961):35-41. *D G T*

Subjects wore strips of reflecting tape on the side of the face, neck, forearm, and thigh; and a camera shutter was opened from the point at which they started to stand up until their weight was off the chair, thereby recording successive light flashes of different colors, which yielded "profiles" when plotted on graphs. The profiles were individually consistent to a high degree. There was also significant difference between the men considered "well-coordinated" and those designated earlier as "poorly coordinated" on the basis of measures of reaction time, rise time, total time, arm pattern, and total path.

479   Jones, Frank P.; Hanson, John A.; Miller, John F., Jr.; and Bossom, Joseph. "Quantitative Analysis of Abnormal Movement: The Sit-to-Stand Pattern." *American Journal of Physical Medicine* 42 (1963):208-218. *D G P T*

Equipment and techniques for obtaining multiple-image photographs with successive images in five colors are clearly described and illustrated. Their use in objectively measuring the paths of different body parts in a complex movement pattern such as standing up is discussed; also demonstrated is how such technique yields graphs and statistical data distinguishing eight neurologically impaired subjects from normals.

480   Jones, Harold E. "The Study of Patterns of Emotional Expression." In *Feelings and Emotions*, edited by M. L. Reymert, pp. 161-168. New York: McGraw-Hill Book Company, 1950. *G* [New York: Hafner Publishing Company]

A report of research indicating that in some children and adolescents a low GSR may occur among those with a great deal of animation and uninhibited overt behavior, whereas high GRS may be found in those with less motor activity and animation.

481   Jones, Harry. *Sign Language.* London: The English Universities Press, 1968. 180 pp. *D*

This well-illustrated manual of deaf sign language as practiced in northern England includes a historical sketch of the one-handed alphabet, beginning with literature from sixteenth-century monastic life.

482   Jones, Ivor H. "Observations on Schizophrenic Sterotypies." *Comprehensive Psychiatry* 6 (1965):323-335. *T*

A phenomenological study of thirteen chronic schizophrenic patients who showed movement stereotypies, defined as one or more of seventeen actions such as rocking, grimacing, standing in a rigid posture, etc. Included is an interesting review of the literature on abnormal movements and schizophrenia, as well as numerous descriptions of patients' actions—how and when they occurred, how patients explained them, and how for a few patients the actions apparently related to delusions. Cessation of medication had different effects on different patients. Some evidence is presented that the stereotypies could be inhibited by inducing attention or getting the patient to engage in another activity and that such actions increased most when someone approached the patient.

483   Jones, Lucian T., Jr. "Children's Galvanic Skin Response and Rated Motor Behavior in Relation to Maternal Authoritarianism." Ph.D. dissertation, University of Houston, 1965. 91 pp. (Datrix order no. 65-9904)

Part of this study presents evidence that children with mothers rated high on authoritarianism got high ratings on "generalized motor activity," compared with children having mothers rated low on authoritarianism.

484   Jones, Marshall R. "Studies in 'Nervous' Movements: I, The Effect of Mental Arithmetic on the Frequency and Patterning of Movements." *Journal of General Psychology* 29 (1943):47-62. *G T*

Observations of which body part moved during periods of rest were made, and arithmetic tests were given. Movements were more frequent during the mental arithmetic problem. Patterns of movement during the rest periods were similar.

485 **Jones, Marshall R.** "Studies in 'Nervous' Movements: II, The Effect of Inhibition of Micturition on the Frequency and Patterning of Movements." *Journal of General Psychology* 29 (1943):303-312. *G T*

A study indicating that drinking a lot of water and waiting increases the number of "genital" and "pedal" movements!

486 **Jourard, Sidney M.** "An Exploratory Study of Body-Accessibility." *British Journal of Social and Clinical Psychology* 5 (1966):221-231. *D T*

Unmarried college students responded to the author's questionnaire inquiring about what parts of other peoples' bodies they had seen and touched and what parts of their own bodies had been seen or touched by others. The results are intriguing. For example, daughters exchange contact on more areas than sons; physical contact between fathers and sons was greatly restricted; Jewish females were touched by boyfriends in fewer areas than Catholic or Protestant girls; and subjects who considered themselves attractive reported being touched in more places. Diagrams of male and female "body accessibility" are presented.

487 **Jourard, Sidney M., and Rubin, Jane E.** "Self-Disclosure and Touching: A Study of Two Modes of Interpersonal Encounter and Their Inter-Relation." *Journal of Humanistic Psychology* 8 (1968):39-48. *D T*

College students answered a questionnaire about whom they touch, with what frequence and where on the body, and who touches them and how often. They also responded to a questionnaire about what they will disclose about themselves to whom. Sons exchange much less contact with fathers. Most contact of both sexes is with their closest opposite-sex friend. There is evidence of consistent individual traits of "touchability." And body contact and self-disclosure were largely independent measures, except for men in relation to male friends and women in relation to male friends.

488 **Kaeppler, Adrienne L.** "Tongan Dance: A Study in Cultural Change." *Ethnomusicology* 14 (1970):266-277.

The author demonstrates how different dances of this Polynesian culture have evolved over a 200-year period, with some having the same form today as those described by Captain James Cook, but others changing under Western influences. The dances are carefully described and related to the social history of the Tongans.

489 **Kagan, Jerome, and Lewis, Michael.** "Studies of Attention in the Human Infant." *Merrill-Palmer Quarterly* 11 (1965):95-127. *G T*

Using heart rate, visual fixation, and body movement as measures of attention in six-month- and thirteen-month-old infants, the authors report response patterns to varying stimuli such as photographs of faces. Sex differences are noted. There appears to be an inverse relationship between sustained attention and amount of bodily activity.

490 **Kalish, Beth.** "The Role of Movement Cues in Observing, Assessing, and Therapeutically Engaging an Autistic Child." Presented at the American Orthopsychiatric Association meeting, San Francisco, 1970. 11 pp. (Copies from author, Developmental Center for Autistic Children, 120 North 48th St., Philadelphia, Pa. 19139)

The author describes a brief movement therapy session with an autistic child and how she observed and moved with the child. A checklist for movement assessment based on Laban's effort-shape analysis is included.

491   **Kalish, Beth.** "A Study of Nonverbal Interaction in the Classroom." Unpublished paper, n.p., 1970. 26 pp. *N T* (Copies from author, Developmental Center for Autistic Children, 120 North 48th St., Philadelphia, Pa. 19139)

The movement patterns of children in five reading groups with the same teacher were observed. On the basis of effort-shape notations of a number of movemnt parameters, the author correctly predicted which was the high-ability and which was the low-ability reading group. A movement comparison of both groups is presented: e.g., the teacher had freer movement and more postural shifts with the higher group. The low readers showed no movement toward peers, low rate of postural shifts, frequent glances at the teacher, and so on.

492   **Kalish, Beth.** "Eric." Unpublished paper, Philadelphia, 1971. 25 pp. *N T* (Copies from author, Developmental Center for Autistic Children, 120 North 48th St., Philadelphia, Pa. 19139)

The developmental and family history, diagnosis and body movement analysis, and treatment of a four-year-old boy with minimal brain damage and autistic features. Included is a description of movement therapy sessions with his mother, as well as behavior ratings and movement assessments of his improvement over one month.

493   **Kalish, Beth.** "Body Movement Scale for Atypical Children." Ph.D. dissertation, Bryn Mawr College, forthcoming. (Copies from author, Developmental Center for Autistic Children, 120 North 48th St., Philadelphia, Pa. 19139)

Includes instructions for the use of a movement rating scale for autistic children, explanation of each item, and description of the movement characteristics of these children. Developed by an experienced movement therapist, the scale has two parts, one for "passive" and one for "active" types of children, with items dealing with energy level, variety and type of movement, orientation to space, etc.

494   **Kanner, Leo.** "Judging Emotions from Facial Expressions." *Psychological Monographs,* Whole No. 186, 41 (1931):1-91. *P T*

This monograph begins with a review of literature from early works on physiognomy and facial anatomy to recognition studies of the 1920's. The author reports a study of recognition of emotions in photographs shown to medical students and evaluates procedures and linguistic and situational aspects of such studies. Individual ability to judge facial expressions is analyzed. Thirty-five pages of data are included.

495   **Kardiner, Abram.** *The Traumatic Neuroses of War.* New York: Paul B. Hoeber, 1941. 258 pp. *T*

Tics and "defensive ceremonials," or ritualized movements, are among the symptoms of traumatic neuroses described and analyzed regarding their origins in war experiences.

496   **Katzell, Raymond A.** "Relations between the Activity of Muscles during Preparatory Set and Subsequent Overt Performance." *Journal of Psychology* 26 (1948):407-436. *P T*

Apparatus for measuring muscle tension in terms of muscle thickness is described. In general, little difference was found between the two phases as measured in this study.

497   **Kealiinohomoku, Joann Wheeler.** "A Comparative Study of Dance as a Constellation of Motor Behaviors among African and United States Negroes." Master's thesis, Northwestern University, 1965.

Although a copy was not obtained for review, this thesis is cited as one of the few sources on the topic.

498   **Kealiinohomoku, Joann W., and Gillis, Frank J.** "Special Bibliography: Gertrude Prokosch Kurath." *Ethnomusicology* 14 (1970):114-128.

A bibliography of the extensive writings of Gertrude Kurath on American Indian dances and ritual, covering from 1946 to 1969.

499   **Keeler, W. R.** "Autistic Patterns and Defective Communication in Blind Children with Retrolental Fibroplasia." In *Psychopathology of Communication*, edited by P. H. Hoch and J. Zubin, pp. 64-83. New York: Grune & Stratton, 1958.

Developmental characteristics and social behavior of five blind children are discussed as similar to those of autistic children and are contrasted with the behavior of normal blind children. Along with an analysis of the etiology of the behaviors, there are descriptions of the movement patterns of the autistic and non-autistic blind children.

500   **Kellogg, W. N.** "The Effect of Emotional Excitement upon Muscular Steadiness." *Journal of Experimental Psychology* 15 (1932):142-166. *G T*

The effects of shock or surprise stimuli on performance of a stylus test. The steadiness of subjects reporting severe induced emotion was markedly disrupted.

501   **Kempf, Edward J.** *Psychopathology.* St. Louis: The C. V. Mosby Company, 1920. 762 pp. *D P*

In this massive work on severe psychopathology, the author draws on experiences with and case studies of schizophrenic patients at St. Elizabeth's Hospital. There is a great deal of attention to disturbances of postures, facial expressions, and body movement, and the illustrations reflect how severe these were in institutionalized mental patients at that time. Within his exposition of the nature, origin, and interpretation of psychopathology, the author includes a surprising number of interpretations of postures and body expressions in works of art.

502   **Kendon, Adam.** "Some Functions of Gaze-Direction in Social Interaction." *Acta Psychologica* 26 (1967):22-63. *N T*

The importance of "the Look," eye-contact, and being looked at is discussed, citing very old references as well as recent research and reporting on an elaborate study of films of dyads in conversation. Five-minute segments from the films were analyzed and recorded with a notation system devised by Exline and Kendon for eye, brow, mouth, and head movement and gaze direction. Data illustrated with figures and tables on how the gaze direction varied according to the point in the utterance and how gaze varied during laughter, interruptions, accompanying remarks of the listener, and short questions are presented, with some notes of head and facial movement. The function of gaze direction (here meaning looking at or away from the other) in regulating emotional arousal, signaling attention, and monitoring the other is discussed.

503   **Kendon, Adam.** "Movement Coordination in Social Interaction: Some Examples Described." *Acta Psychologica* 32 (1970):100-125. *D N*

Corroborating W. S. Condon's observations of interactional synchrony, Kendon presents very detailed examples with diagrams of how people "keep pace" with each other in interaction, move synchronously together, and coordinate their postural shifts and actions apparently according to the degree of their "presence" or involvement. On the basis of minute examination of several films, sometimes frame by frame, Kendon shows how changes in the direction of each participant's movement can exactly coincide, how changes in the flow of speech often correspond with changes in direction, and how speaker and listener may "mirror" each other's actions.

504   **Kendon, Adam.** "Some Relationships between Body Motion and Speech: An Analysis of an Example." In *Studies in Dyadic Communication*, edited by A. W. Siegman and B. Pope. Elmsford, N.Y.: Pergamon Press, forthcoming. *D T*

This is an important initial effort at analyzing the organization and patterning of gesticulations and positions and how these correspond to speech organization. Kendon defines six units of speech, from the smallest (in this case, the syllable) to the largest (called here "discourse") and demonstrates from a detailed analysis of one film segment how the speaker's gesticulating is also hierarchically organized in units which coincide with these speech units. Notably, the larger the speech unit, the larger (in terms of body part) the preparatory movement immediately preceding it.

505   **Kendon, Adam.** "The Role of Visible Behaviour in the Organization of Social Interaction." In *Expressive Movement and Nonverbal Communication*, edited by M. von Cranach and I. Vine. London: Academic Press, forthcoming.

A well-written paper integrating much of what is so far documented about the visible aspects of conversational behavior and how participants actually organize and maintain a face-to-face encounter. The author focuses particularly on kinesics, proxemics, and ethological research and includes numerous descriptive examples and illustrations from his own work. The exposition ranges from aspects of territoriality, spatial arrangement, and proxemics of the gathering to properties of body orientation, specific positions or postures, head movement, and gaze as these relate to social role, subgrouping, speaker-listener relationship, etc. Included is a discussion of how the movement patterns may be organized and how phrases of speech and body movement can be delineated.

506   **Kendon, Adam, and Cook, Mark.** "The Consistency of Gaze Patterns in Social Interaction." *British Journal of Psychology* 60 (1969):481-494. *T*

After a concise and useful summary of research on eye contact and gaze direction in social interaction, the authors describe a study of gaze and speech patterns in which eleven subjects spoke individually with each of four other subjects for thirty minutes. Recorders behind a one-way screen noted when and for how long S_1 looked at S_2, S_2 looked at S_1, and when and for how long each spoke. Each subject later filled out the MMPI and FIRO tests and was given field dependence-independence tests. Using comprehensive statistical analyses of the data, the authors obtained significant· findings regarding gaze and speech patterns and the identity and sex of subject and partner, interrelations between S_1 and S_2's performances, and gaze patterns and personality data—most notably, that there is marked individual consistency in gaze behavior.

507   **Kennard, David W., and Glaser, Gilbert H.** "An Analysis of Eyelid Movements." *Journal of Nervous and Mental Disease* 139 (1964):31-48. *D G P*

Description of apparatus for measuring upper- and lower-eyelid movements and skin resistance while controlling gaze direction. With data in the form of wave recordings and graphs obtained from thirteen male subjects, the authors report various phenomena such as "eyelid noise," relationship of upper and lower lids, effects of various eye movements on lids, and blinking patterns, particularly in relation to level of attention or wakefulness.

508   **Kessen, William; Hendry, Louise S.; and Leutzendorff, Anne-Marie.** "Measurement of Movement in the Human Newborn: A New Technique." *Child Development* 32 (1961):95-105. *D G T*

Description of equipment and techniques developed to objectively study neonate movement. This included an Esterline-Angus operations recorder and movie camera (with a mechanism to synchronize them) and a film analyzer that makes permanent copies of selected frames, which when projected on tracing paper were analyzed for changes in placement of body parts. The method in their study showed individual differences, an overall increase in activity, and no clear variations in laterality in the first five days of life.

509   **Kestenberg, Judith S.** "The Role of Movement Patterns in Development: I, Rhythms of Movement." *Psychoanalytic Quarterly* 34 (1965):1-36. *G* (Copies from Dance Notation Bureau, 8 East 12th St., New York 10003)

The author hypothesizes that there are congenital motor rhythms observable in early infancy which may be expressive of oral, anal, phallic, etc., drives. She describes the body movement rhythms and activity styles of three children from early infancy to nine months and their later development up to ten years of age, indicating that there are individual consistencies in movement patterns over time. Some predictions made at nine months concerning their temperament, preferred defenses, object relationships, etc., were borne out later, but others were not. Descriptions of the movement and activity of the children at home illustrate their "preferred" rhythms, clashes in rhythms between mother and child, early evidence of disturbance, and how environmental influences appeared to modify the children's original rhythms.

510   **Kestenberg, Judith S.** "The Role of Movement Patterns in Development: II, Flow of Tension and Effort." *Psychoanalytic Quarterly* 34 (1965):517-563. *G* (Copies from Dance Notation Bureau, 8 East 12th St., New York 10003)

A theoretical formulation with respect to the transition from early motor rhythms reflecting drive discharge to forms of regulation observable in the "flow of tension." Flow of tension (the "relation between contractions of agonistic and antagonistic muscles" described as variations in "free" or "bound") is described, and patterns of flow are classified into rhythms corresponding to psychosexual levels and rhythms of body functions. The kinds of flow patterns and how they vary in intensity, duration, etc., are correlated with various affects. It is hypothesized that newborn infants have individually consistent flow patterns and that within these can be seen the "kernels" for later modes of "flow regulation," as well as personality characteristics. Using observations of three children made over ten years, there is extensive discussion of modes of flow regulation in body movement and their behavioral significance; mother-child patterns of movement; and aspects of psychosexual development and motor rhythms, followed by a section on how specific "effort" elements evolve from different flow patterns and their relationship to ego development. This highly technical paper is difficult to understand without actually seeing what it is in the movement that the author is referring to.

511   **Kestenberg, Judith S.** "Rhythm and Organization in Obsessive-Compulsive Development." *International Journal of Psycho-Analysis* 47 (1966):151-159.

Using Freud's analysis of the "Rat Man," and observations of a patient in analysis and of infants, the author discusses two phases of anality in infant development (anal-erotic and anal-sadistic) in terms of movement rhythms (sustained tension and release, holding and dropping), as well as styles of thinking and interacting.

512 **Kestenberg, Judith S.** "Suggestions for Diagnostic and Therapeutic Procedures in Movement Therapy." Presented at American Dance Therapy Association Second Annual Conference, pp. 5-16, Washington, D.C., 1967. (Copies from: ADTA, 5173 Phantom Court, Columbia, Md. 21043)

The author proposes a procedure for obtaining "movement profiles" of children based on the effort-shape analysis of Rudolf Laban and Warren Lamb and additions and refinements made by the author for use in personality assessment, psychodiagnosis, and, if indicated, a treatment plan involving movement reeducation or therapy. After describing the system for analyzing movement, a movement profile of a four-year-old girl with psychological interpretation and treatment recommendations is presented.

513 **Kestenberg, Judith S.** "The Role of Movement Patterns in Development: III, The Control of Shape." *Psychoanalytic Quarterly* 36 (1967):356-409. (Copies from Dance Notation Bureau, 8 East 12th St., New York 10003)

Having discussed "tension flow" in a previous paper as aiding drive differentiation, the author focuses on "shape flow" (movement patterns of "growing and shrinking") and spatial patterns as related to differentiation of self and objects and development of object relations. Integrating concepts and theories developed by Rudolf Laban and Warren Lamb with her own longtitudinal studies of movement and with psychoanalytic theory, Kestenberg discusses the psychological significance of the development of various spatial characteristics: e.g., the relation between the infant's horizontal movements and the oral phase, the vertical plane and the anal phase, the sagittal plane and the urethral phase. This is a comprehensive theory about movement, personality, and development. Individual movement characteristics and their modification, particularly in mother-child interaction, and the movement characteristics of infancy, early childhood, latency, adolescence, and adulthood are described and illustrated.

514 **Kestenberg, Judith S.; Marcus, Hershey; Robbins, Esther; Berlowe, J.; and Buelte, A.** "Development of the Young Child as Expressed through Bodily Movement: I." *Journal of the American Psychoanalytic Association,* in press.

The authors trace development from the neonatal phase through age three in terms of movement patterns of "tension flow," "shape flow," and spatial "dimensional factors," hypothesizing that there are characteristic movement rhythms, spatial planes, and modes of interaction and activity for the oral, anal, and urethral psychosexual stages of development. The significance of these patterns for intrapsychic development, ego organization, and the body ego are discussed, with numerous examples of body movement patterns, mother-child interaction, and activity.

515 **King, H. E.** *Psychomotor Aspects of Mental Disease: An Experimental Study.* Cambridge, Mass.: Harvard University Press, 1954. 185 pp. *D G T*

A series of tests of reaction time, manual dexterity, and tapping rate were given to groups of normal subjects, chronic schizophrenics, and patients with "subacute behavior disorders." The rationale, equipment, procedure, population characteristics, and test results are presented. Marked differences in performance between the groups are described, but very little difference was found in motor learning patterns. The author includes a summary of the study, a theoretical chapter, and a review of experimental research on psychomotor aspects of mental disease.

516 **Kishimoto, Suehiko.** "A Brief Note on the Language of Gesture." *Memoirs of*

Osaka Gakugei University, No. 13 (1964):62-67. (Copies from author, c/o Osaka Kyoiku University, 103-3, Ikeda City, Osaka, Japan).

The author's preliminary research on formalized Japanese gestures and their meaning, including symbolic gestures, formal greeting behavior, dance and religious rituals, sign language, and Japanese words that have an action basis—e.g., "to feel small in Japanese is a word combining 'shoulder,' 'body,' and 'narrow'."

517 **Kleck, Robert E.** "Physical Stigma and Nonverbal Cues Emitted in Face-to-Face Interaction." *Human Relations* 21 (1968):19-28. *T*

In a carefully controlled study using "confederates" who were made to look physically disabled (with one leg amputated), the author found that there was no decrease in the amount of eye contact between the "disabled" person and the subject, but there was a significant decrease in the amount of body movement of the subject as compared with when he spoke to a non disabled "confederate."

518 **Kleck, Robert E.** "Interaction Distance and Non-Verbal Agreeing Responses." *British Journal of Social and Clinical Psychology* 9 (1970):180-182.

Subjects showed more head nods of agreement and self-manipulations when listening to a confederate who sat four feet away than when he sat ten feet away.

519 **Kleck, Robert E., and Nuessle, William.** "Congruence between the Indicative and Communicative Functions of Eye Contact in Interpersonal Relations." *British Journal of Social and Clinical Psychology* 7 (1968):241-246. *T*

The authors consider the relationship between what eye contact reflects about the sender (indicative) and what it communicates to the observer (communicative). Films of a confederate showing either high or low eye contact were judged by students. They judged that the high-eye-contact subject liked the interviewer more and was more comfortable. Also, the female students judged the amount of eye contact better than the men.

520 **Kline, Linus W., and Johannsen, Dorothea F.** "Comparative Role of the Face and of the Face-Body-Hands as Aids in Identifying Emotions." *Journal of Abnormal and Social Psychology* 29 (1935):415-436. *P T*

Groups judged the emotion expressed in photographs of either the face alone or the face, trunk, and arms. In one experiment judges used their own terms; in another they selected terms from a list. A marked practice effect was noted, with body-face pictures yielding better scores in some ways and the supplying of words improving judgments 16 percent. Considering this and the stereotyped terms given, the author concludes that expression and recognition of emotion is not well developed in modern life.

521 **Klineberg, Otto.** "Racial Differences in Speed and Accuracy." *Journal of Abnormal and Social Psychology* 22 (1927):273-277.

Children of the same town, half of them white and half Yakima American Indian, showed marked differences in the manner in which they performed structured tests. The Indian children took more time and were more accurate; the whites were more hurried and impulsive.

522 **Klineberg, Otto.** "Emotional Expression in Chinese Literature." *Journal of Abnormal and Social Psychology* 33 (1938):517-520.

In preparation for a study of expression in the Chinese, the author reviewed pertinent Chinese literature—a few prominent novels and books on etiquette, ritual, and acting—and enumerated characteristic facial expressions and gestures and their meaning, which he deduced from the writings.

523   **Knapp, Robert H.** "The Language of Postural Interpretation." *Journal of Social Psychology* 67 (1965):371-377. *T*

Subjects were asked to judge the interaction between silhouette figures of a man and woman placed in various positions. Depending on the sex of the figure, its level and rotation, its head position, and its relation to the other figure, consistent judgments about who was "initiating" and whether they were "telling" or "demanding" could be made.

524   **Knight, Robert P.** "Psychotherapy of an Adolescent Catatonic Schizophrenia with Mutism." *Psychiatry* 9 (1946):323-339.

A detailed account of successful psychotherapy with a catatonic young man, including a summary of the patient's history, much of the actual verbal and non-verbal behavior of patient and therapist, and a summary interpretation by the therapist. Beautifully written and full of observations linking the patient's actions or moments of catatonic withdrawal to the context and specific verbal themes, this paper provides insight into the nature of catatonic rigidity.

525   **Koch, Helen L.** "An Analysis of Certain Forms of So-Called 'Nervous-Habits' in Young Children." *Journal of Genetic Psychology* 46 (1935):139-170. *T*

A study of eleven types of body-directed movements observed in forty-six nursery children. Differences related to sex, degree of physical restraint, personality characteristics, and number of friends are noted. Few correlations were found between the mannerisms and parental characteristics, I.Q., or age.

526   **Kogan, Kate L., and Wimberger, Herbert C.** "An Approach to Defining Mother-Child Interaction Styles." *Perceptual and Motor Skills* 23 (1966):1171-1177. *T*

Techniques for recording the nonverbal interaction of mother and child in a semistructured playroom situation are outlined, and preliminary data on differences between a few mother-child pairs from different socioeconomic backgrounds are presented. The system is not fully described but includes coding individual nonverbal behaviors and categorizing modes of interaction as verbal, visual, tactile, manipulative, and/or gestural.

527   **Krapf, E. Eduardo.** "Transference and Motility." *Psychoanalytic Quarterly* 26 (1957):519-526.

A discussion of the defensive function of body movements and muscle tension observed in psychoanalysis, with case examples and theories about the origin of the motor patterns in intrauterine and early childhood experience.

528   **Kreezer, George.** "Motor Studies of the Mentally Deficient: Quantitative Methods at Various Levels of Integration." *Training School Bulletin* 32 (1935): 125-135. *G T*

A pilot study is outlined in which various levels of motor functioning, ranging from a complex motor activity such as walking to irritability of the muscles, are assessed by different methods (from film analysis to the "chronaxy technique"). Differences between normal and feebleminded subjects in amplitude of movement and excitability of muscles are reported.

529   **Kretschmer, E.** *Physique and Character: An Investigation of the Nature of Constitution and of the Theory of Temperament.* Translated by W. J. H. Sprott. New York: Harcourt, Brace and Company, 1925. 266 pp. *P T* [New York: Cooper Square Publishers]

Kretschmer's analysis of body types as presented here is complex and detailed in spite of his defining general types (asthenic, athletic, and pyknic). He concentrates on physique characteristics of schizophrenic and manic-depressive patients and gives considerable attention to their muscle tonus and movement characteristics.

530 Krim, Alain. "A Study in Non-Verbal Communications: Expressive Movements During Interviews." *Smith College Studies in Social Work* 24 (1953):41-80. *T*

The movement behavior of six male patients was observed during therapy sessions behind a one-way screen, and the observations were compared with case summaries, physical characteristics, therapist's comments, and verbal content. Samples of the movement descriptions and accompanying verbal content, together with interpretations, are presented.

531 Kris, Ernst. "Laughter as an Expressive Process." *International Journal of Psycho-Analysis* 21 (1940):314-341.

More than a psychoanalysis of the origins and nature of laughter, this is a psychoanalytic discussion of expressive movement, its differentiation in development, its regulation by the ego, the conditions under which one may return to "the archaic type of expression," total body movement, and so on. There is considerable discussion of the nature of smiling, with description of a man who suffered from compulsive laughter as an example of its relation to anxiety. There is also an analysis of the disturbances of regulation and timing in the expressive behavior of schizophrenic patients.

532 Kris, Ernst. "Some Comments and Observations on Early Autoerotic Activities." *Psychoanalytic Study of the Child* 6 (1951):95-116.

A psychoanalytic discussion with case examples and behavioral observations of normal and disturbed children. The role of autoerotic and rhythmic activity in promoting and, under certain conditions, impeding development is discussed.

533 Krout, Maurice H. "A Preliminary Note on Some Obscure Symbolic Muscular Responses of Diagnostic Value in the Study of Normal Subjects." *American Journal of Psychiatry* 88 (1931):29-71. *T*

Reading this excellent examination of early literature on movement behavior, one has a sense of how productive and serious body movement research promised to be in 1931, only to fade in the 1940's and 1950's. Krout reviews four areas: experimental study of reflex movements, particularly Pavlov's animal studies and Watson's lists of neonate movements; the symbolism and psychodynamic origins of "chance acts," postures, and tics according to the psychoanalytic view; the psychological research on facial or "mimetic" expression of emotion; and the objective measurement of gestures (Olson) and analysis of muscle tension (Jacobsen). He proposes that study of automatic, fleeting responses done by normal people is greatly neglected; presents a long list of movements he has observed in everyday life; and discusses problems of classification of nonverbal behavior. He then illustrates a discussion of the nature of symbolism in gestures with observations of actions a student made coincident with specific remarks by the professor, and outlines a nonverbal research project using observation of students in the classroom.

534 Krout, Maurice H. "Symbolic Gestures in the Clinical Study of Personality." *Transactions of the Illinois State Academy of Science* 24 (1931):519-523.

A brief paper on experimental versus clinical assessment of personality and Krout's modified experimental techniques for studying an individual's gestures.

535   Krout, Maurice H. "Autistic Gestures: An Experimental Study in Symbolic Movement." *Psychological Monographs,* Whole no. 208, 46 (935):1-126. *G T*

Krout reports on a number of studies of "autistic gestures" performed by students in the classroom, in experimental word-association tasks, in interviews, and in introspective exercises. Students' "modal interpretations" of 160 gestures and subjects' associations to gestures performed during reverie and conversation are presented. Difficulties in and reliability of observing students' gestures in the classroom, and later as they accompanied word associations, are analyzed. Consistency of the gesture according to situation is assessed. In another study, gestures and words produced during an interview were duplicated exactly by a subject under trance. Krout concludes with a series of association experiments and a study of two individuals indicating a relationship between autistic gestures and emotional conflict and tension.

536   Krout, Maurice H. "The Social and Psychological Significance of Gestures: A Differential Analysis." *Journal of Genetic Psychology* 47 (1935):385-412. *T*

A theoretical paper on the origins, development, and nature of four groups of gestures defined by Krout: nonsocial, conventional, pseudo-conventional, and autistic.

537   Krout, Maurice H. "Further Studies on the Relation of Personality and Gesture: A Nosological Analysis of Autistic Gestures." *Journal of Experimental Psychology* 20 (1937):279-287. *T*

Analyses of observations of movements of students in a classroom and an experimental situation. The gestures are classified according to which area of the body moves and the "goal" of the movement (anatomical parts touched, "environmental" goals, etc.). A table of frequency of the different actions for each individual in each situation is presented. Statistical analysis shows high individual consistency, with indications of gesture types and special changes under stress for some subjects.

538   Krout, Maurice H. "An Experimental Attempt to Produce Unconscious Manual Symbolic Movements." *Journal of General Psychology* 51 (1954):93-120. *D T*

A study of "autistic" gestures, which in Krout's formulation are a form of self-communication and tension reduction. After some unpublished research on judgment of hand gestures from newspaper photographs and experimental attempts to induce such gestures is cited, the rationale and procedures for a series of experiments are reported. Each of 100 selected subjects was confronted with a set of fifteen conflicting statements and directed to answer the first immediately and to wait until cued after the second. Observers noted his gestures, position, etc. The series was then repeated, and this time the subject was to note his attitude when first presented the statements. The procedures, observer training, protocol questions, and what emotion each statement was designed to elicit are described.

539   Krout, Maurice H. "An Experimental Attempt to Determine the Significance of Unconscious Manual Symbolic Movements." *Journal of General Psychology* 51 (1954):121-152. *T*

The results of the experiment just cited are reported in this article. Reliability of observation of the gestures was not sought; the focus here was on validity. Subjects' attitudes on later repetition of the questioning proved consistent. Also the subjects' reports corroborated what the experimenter predicted their attitudes would be in response to conflicting statements. Statistical analysis yielded the number of significant gestures and attitudes or emotions and which gestures consistently occurred with which attitudes. Ten gestures correlated significantly with specific attitudes: e.g., fist gestures with aggression, hand to nose with fear, finger on lips with shame, etc. In addition, marked sex differences in the "autistic" manual gestures but not in the attitudes were found, with eight predominatly male, nine female, and fifteen shared gestures listed.

540 **Kulka, Anna M.; Walter, Richard D.; and Fry Carol P.** "Mother-Infant Interaction as Measured by Simultaneous Recording of Physiological Processes." *Journal of the American Academy of Child Psychiatry* 5 (1966):496-503. *G*

Twelve mothers and their one- to three-month-old infants were filmed, and EKG and EMG recordings were made while they nursed, interrupted, and then resumed nursing. EMG and EKG recordings increased during the interruption for most of the mothers and infants. Behavioral observations of mothers with high EMG recordings are noted.

541 **Kunst, Mary S.** "A Study of Thumb- and Finger-Sucking in Infants." *Psychological Monographs*, Whole no. 290, 62 (1948):1-71. *G T*

A most comprehensive study of finger-sucking in 143 infants from one to fifty-two weeks old and its relation to teething, feeding schedules, amount and type of diet, presence or absence of adults, body position, age, sex, and individual differences.

542 **Kurath, Gertrude Prokosch.** "A New Method of Choreographic Notation." *American Anthropologist* 52 (1950):120-123. *N*

The author sketches a relatively simple dance notation graphically recording the "ground plan, the steps and gestures, and they rhythmic beat and structure" as applied to Mexican and Indian dances.

543 **Kurath, Gertrude Prokosch.** "Panorama of Dance Ethnology." *Current Anthropology* 1 (1960):233-254. *D N*

A valuable bibliography and review of literature on dance and culture. Literature on specific areas is cited: cultural differences in individual or group stress and male-female roles in the dance, "choreographic areas" of the world, rituals, patterns of transculturation, decline and resurgence in dance, and brief notes on dance in relation to such fields as psychology and linguistics. There is a discussion of the use of movement notations, particularly Labanotation, for dance ethnology and the problems of training dance ethnologists.

544 **Kurath, Gertrude Prokosch, and Marti, Samuel.** *Dances of Anáhuac: The Choreography and Music of Precortesian Dances.* Chicago: Aldine Publishing Company, 1964. 251 pp. *D N P*

The dances and rituals of the ancient Aztecs are reconstructed from their art, notated in Labanotation, and illustrated with photographs and drawings. There are chapters on the postures, gestures, and steps of the dances. The relationship between the dance and life of the Aztecs is discussed throughout this impressive book.

545 **Kwint, L.** "Ontogeny of Motility of the Face." *Child Development* 5 (1934): 1-12. *T*

In this experimental study 476 children, ages four to sixteen, were asked to perform a number of facial movements such as "lifting the brows," screwing up one eye," "drawing down the lower lip," etc. The results show a clear developmental trend in "mimic psycho-motility," with retarded children performing far fewer movements than normal children. A graded scale of which movements were possible at each age is presented.

546 Laban, Rudolf. *Principles of Dance and Movement Notation.* London:Macdonald & Evans, 1956. 56 pp. *N*

Primarily a textbook presenting the principles and theoretical basis of Labanotation (or Kinetography Laban), there are numerous examples of body movements, mostly from dance, notated with this system. The book is particularly interesting for Laban's discussion of how the system evolved; what he sees as the problems in developing a "language" and efficient method of notating movement; how his notation has been applied; and the potential he sees for the notation in behavioral research, rehabilitation, and therapy as well as in dance.

547 Laban, Rudolf. *The Mastery of Movement.* Edited by L. Ullmann (1950; 2d ed., rev., London: Macdonald & Evans, 1960). 186 pp. *N P T*

In his lifetime Rudolf Laban, an important European choreographer and inventor of a dance notation, explored movement in relation to drama, dance, mime, work skills, culture, and personality. Although this book, first entitled *Mastery of Movement on the Stage*, is especially addressed to the performing artist, it is a very important theoretical work relevant to many disciplines. In chapters such as "The Significance of Movement," "The Roots of Mime," and "The Study of Movement Expression," developmental, cultural, and psychological as well as artistic aspects of all types of movement are discussed. Laban's movement terminology and concepts are presented in a series of exercises for the reader to perform, and specifically in charts of "effort" terms and their psychological interpretation. At times the extravagant, overextended writing belies how concrete and disciplined Laban's analysis can be; but this book furnishes a wealth of observations and hypotheses about movement behavior, particularly the relationship between movement and emotion.

548 Laban, Rudolf. *Modern Educational Dance.* 2d rev. ed. Revised by L. Ullmann. London: Macdonald & Evans, 1963. 114 pp. *N* [New York: Praeger Publishers, 1968]

A concise presentation of the early form of Laban's "effort" analysis and notation for body movement quality and spatial patterns and his concepts of kinesphere, personal space, and types of pathways. Included are an analysis of combinations and patterns of "effort" qualities or types of movement intensity, with notes on developmental stages in children's movement.

549 Laban, Rudolf. *Choreutics.* Annotated and edited by L. Ullmann. London: Macdonald & Evans, 1966. 214 pp. *D N*

Laban introduces his principles of "space harmony" and the aesthetics and dynamics of spatial patterns of movement in an approach to body movement and dance theory that is comparable to music theory and the study of harmony. Given one's kinesphere or "reach space," three planes of space, and the myriad combinations and sequences of spatial directions, Laban explores various "trace-forms" and "scales," their harmonic relationships, rhythmic properties, and qualitative effects.

550   **Laban, Rudolf, and Lawrence, F. C.** *Effort.* London: Macdonald & Evans, 1947. 88 pp. *D N* [New York: International Publications Service, 1971]

Although it is primarily addressed to work efficiency and analysis of work actions, psychological aspects of movement are discussed in this presentation of Laban's system for analyzing rhythm and body movement dynamics (which later evolved into his "effort-shape analysis"). Combinations and sequences of effort dynamics, considered space, force, time, and flow variations are analyzed and notated. Emotional and personality aspects of effort patterns are noted.

551   **La Barre, Weston.** "The Cultural Basis of Emotions and Gestures." *Journal of Personality* 16 (1947):49-68.

With wit and numerous examples, the author describes how differently people around the world may gesture "yes" or "no," point, walk, give or take things, express emotion, greet each other, indicate approval or disapproval, show affection, and so on. Arguing that motor patterns are largely learned and culturally determined, La Barre continues with a comparison of sign languages, styles of acting, and types of hand gestures.

552   **Lamb, Warren.** *Posture and Gesture: An Introduction to the Study of Physical Behaviour.* London: Gerald Duckworth & Co., 1965. 189 pp. *D N*

An introduction to effort-shape analysis of movement as developed and used by the author for assessing individual body movement patterns and aptitude. Examples of how Lamb diagrams, notates, and interprets patterns of movement quality or dynamics are presented.

553   **Lamb, Warren, and Turner, David.** *Management Behaviour.* New York: International Universities Press, 1969. 177 pp. *D*

Although the authors also explore general issues of management and business administration, the major focus of the book is unique: using analysis of individual body movement styles for assessment of managers and executives in British industry. The method of observing movement and abstracting individual movement profiles based on the "effort" concepts originated by Laban is presented in some detail. How the profiles are interpreted with regard to personality characteristics, styles of relating to colleagues, decisionmaking potential, etc., is discussed, and some case studies of managers are presented. Whereas the authors state that the movement assessment technique has been used in management consulting some several thousand times over a number of years with marked success, controlled research establishing its reliability and validity is not cited.

554   **La Meri.** *Dance as an Art-Form: Its History and Development.* New York: A. S. Barnes and Co., 1933. 198 pp. *T*

Particularly relevant here are the chapters on the styles of Eastern, (Chinese, Japanese, Indian, Arabic, etc.), European (Spanish, Slavic, Scandinavian, etc.), and South American dances, described and interpreted. Included is a table of over 100 dances discussed in the text, citing where they are from and supplying notes on their origins, group form, and style.

555   **La Meri.** *The Gesture Language of the Hindu Dance.* (1941; reprint ed., New York: Benjamin Blom, 1964). 100 pp. *P*

Following a summary of differences between Occidental and Oriental dance and of the historical and cultural background of Indian dance, numerous photographs of Hindu dance gestures and descriptions of their meaning are presented.

556 **La Meri.** *Spanish Dancing.* Pittsfield, Mass.: Eagle Printing and Binding Company, 1967. 157 pp. *P*

Includes a history of flamenco and other Spanish dances, description of regional dances, and a chapter on technique.

557 **Landis, Carney.** "Studies of Emotional Reactions: I, 'A Preliminary Study of Facial Expression'." *Journal of Experimental Psychology* 7 (1924):325-341. *P*

Nineteen male subjects were filmed during a series of situations such as listening to classical music, looking at nudes, smelling ammonia, etc. The expected expressive responses to these are described and illustrated, but a number of subjects showed different facial and body expressions, or hardly any at all. When subjects imitated an expression, it was often a traditional one, and not what they spontaneously did. Landis argues that, although expressions of "primary emotions" may have a phylogenetic basis, the subtler ones appear culturally learned.

558 **Landis, Carney.** "Studies of Emotional Reactions: II, General Behavior and Facial Expression." *Journal of Comparative Psychology* 4 (1924):447-509. *D P T*

Noting that until that time it had not been done, Landis did an elaborate study of the actual behavior and facial expressions of individuals in the same emotionally arousing situation. Films were made of subjects looking at pictures of skin diseases, reading pornographic histories, being shocked, etc. Series of photographs were made of the various facial expressions displayed under each condition, and these were subjected to elaborate analysis and measurement (amount of involvement of different muscles, head positions, etc.) and correlated with the situation or verbal reactions. No facial expression was characteristic of any situation or verbalization; however, individual consistencies in facial expression were noted. There are also descriptions of what body movements were typically observed in each situation and measures of the degree of expressiveness, with men rating higher than women. Landis discusses the significance of his findings and their contradiction of much previous work.

559 **Landis, Carney.** "The Interpretation of Facial Expression of Emotion." *Journal of General Psychology* 2 (1929):59-72. *T*

Using photographs of faces from a previous experiment of subjects in emotional contexts or attempting to portray a given emotion, the author had observers judge the emotion expressed and the situation that elicited it. He concluded from the results that observers could not predict either the emotion or the context better than by chance.

560 **Landis, Carney.** "Emotion: II, The Expressions of Emotion." In *A Handbook of General Experimental Psychology,* "Part I: Adjustive Processes," edited by C. Murchison, pp. 312-351. (Clark University Press, 1934; reprint ed., New York: Russell & Russell, Publishers, 1969).

The author reviews Darwin's, Wundt's, and James's theories about emotion and expressive body movements, and then surveys early physiological treatises on facial expression, experiments on recognition of facial expression, and developmental studies of smiling and crying. He concludes with a review of research on physiological changes in emotion.

561 **Landis, Carney; Giullette, Ruth; and Jacobsen, Carlyle.** "Criteria of Emotionality." *Journal of Genetic Psychology* 32 (1925):209-234. *G T*

Twenty-five subjects were given a series of physiological and psychological tests, and their reactions to a series of emotional situations were photographed to determine the best criteria of "emotionality" and of "emotional stability." Ratings of their stability and expressiveness were made by people who knew them and were compared with objective measurements. The following were found to be good measures of expressiveness: head movements, amount of laughter, increased reaction time, and the Woodworth questionnaire. Range and variability of blood pressure and the speed of tapping were not judged reliable.

562 **Landis, Carney, and Hunt, William A**. "Studies of the Startle Pattern: III, Facial Pattern." *Journal of Psychology* 2 (1936):215-219. *P T*

Film analysis of the changes in facial expression of twenty-nine subjects during a gunshot, noting the characteristic pattern and individual variation.

563 **Lange, Carl G., and James, William.** *The Emotions.* Vol. I. Baltimore: The Williams & Wilkins Company, 1922. 135 pp. [New York: Hafner Publishing Company]

Reprints of articles on the nature of emotion, written independently by the two authors but containing what came to be called the James-Lange Theory: namely, that emotions are the experience of bodily sensations and movements which are themselves the immediate response to perceptions of exciting stimuli. There is extensive duscussion of the role of muscle tension and movement in emotional states.

564 **Langer, Susanne K.** "Virtual Powers." In *Feeling and Form: A Theory of Art,* pp. 169-187. New York: Charles Scribner's Sons, 1953.

As part of a comprehensive and important aesthetic theory, the author discusses the nature of dance, its uniqueness and independence from other arts, and its roots in gesture and *"imagined* feeling in its appropriate physical form." Dance is seen as the interplay of tensions and forces, the portrayal of the subjective experiences of human volition, vitality, and the power to perceive, comprehend, and create changes. The expressive gesture, as both symbolic and symptomatic of a subjective condition, is discussed.

565 **Langer, Susanne K.** "The Magic Circle." In *Feeling and Form: A Theory of Art,* pp. 188-207. New York: Charles Scribner's Sons, 1953.

Discussing the function of dance in primitive groups as portraying vital and mysterious forces, Langer focuses on the circle formation as delineating the holy realm and separating it from the profane world outside, and on the transformation from the sacred content and function of the dance to its secular forms. Of particular note in this profound and important writing is the discussion of why dance is not primarily an art of space, time, drama, or poetry but of "virtual powers," as well as the relationship between the visual and kinesthetic senses and communication in dance.

566 **Langfeld, Herbert S.** "The Judgement of Emotions from Facial Expressions." *Journal of Abnormal and Social Psychology* 13 (1918):172-184. *D*

An early experiment on judgment of facial expression using artists' drawings. Subjects wrote down their interpretations; were then shown pictures again to test response constancy; and later were shown all the interpretations plus those accompanying the pictures. No statistical results are presented, but the author states that judgment was very good, with many good approximations of the picture title.

567   **Langworthy, Orthello R.** "The Differentiation of Behaviour Patterns in the Foetus and Infant." *Brain* 55 (1932):265-277.

Description of the movements and reflexes of animal and human foetuses, and their significance and differentiation over time.

568   **Langworthy, Orthello R.** *The Sensory Control of Posture and Movement: A Review of the Studies of Derek Denny-Brown.* Baltimore: The Williams & Wilkins Company, 1970. 145 pp. *D*

In this review and synthesis of Denny-Brown's work, there are numerous examples of how specific brain lesions (in animals and in humans) produce specific motor disturbances. This yields a wealth of hypotheses concerning the loci or neurophysiological bases of specific movement patterns and characteristics. Much more than a neurophysiological view of posture or reflexes, this work offers rich and provocative observations and analyses of many aspects of movement behavior.

569   **Lanzetta, John T., and Kleck, Robert E.** "Encoding and Decoding of Nonverbal Affect in Humans." *Journal of Personality and Social Psychology* 16 (1970):12-19. *T*

Subjects had to judge whether those on a videotape were seeing a red light (signaling the onset of shock) or a green light (no shock), a situation they themselves had experienced. Judgments were better than chance, although the authors report that punishing the judges' errors with shock did not improve their recognitions. Notably, subjects easiest to judge made more errors of judgment.

570   **La Piere, Richard T., and Farnsworth, Paul R.** *Social Psychology.* New York: McGraw-Hill Book Company, 1949. 626 pp.

Includes a critique of literature on physiognomy and the relation between body physique and personality and a brief chapter on gesture and person perception, communication, cultural differences, and mother-infant interaction.

571   **Lassner, Rudolf.** "Annotated Bibliography on the Oseretsky Tests of Motor Proficiency." *Journal of Consulting Psychology* 12 (1948):37-46.

Forty-four works are reviewed, many of them Russian. There is also a brief survey of other motor development tests and mental ability.

572   **Lavater, John C.** *Essays on Physiognomy: Designed to Promote the Knowledge and the Love of Mankind.* 8th rev. ed. Translated by T. Holcroft. London: William Tegg and Company, 1853. 507 pp. *D N* [Farnborough, Eng.: Gregg International Publishers, 1971]

Possibly the best work on physiognomy or judgment of character from the form and expression of head, face, and body. Lavater analyzes individuals from Socrates to Attila, national differences, family resemblances, sex differences, and the character of animals as well as children and adults. The following passage illustrates how perceptive observations of body expression and physique are intermingled with myriad prejudices and myths of the time in this book: "He who walks slowly, bending forwards; who retreats in advancing to meet thee; who says gross and rude things in low and timid voice; who fixes his eyes on thee so soon as thou hast turned from him, and never can look thee calmly and steadily in the face; . . . Oh! couldst thou feel his skull! what concealed misconformation, what irregular knots, what parchment softness, and at the same time hardness wouldst thou find!—Avoid him. . . ." (p. 488)

573 **Lawler, Lillian B.** *The Dance in Ancient Greece.* Middletown, Conn.: Wesleyan University Press, 1964. 160 pp. *D P*

An excellent history of ancient Greek dance and its relation to Greek culture, ritual, and drama from prehistoric Crete through the Classical period. Richly illustrated, the book appears based on extensive study of Greek art and literature.

574 **Lebowitz, Martin H.; Colbert, Edward G.; and Palmer, James O.** "Schizophrenia in Children: The Frequency and Quality of Certain Motor Acts in Diagnosis." *American Journal of Diseases in Children* 102 (1961):25-27.

Three groups of grammar-school-age children were observed during free play: regressed schizophrenic children, normal children, and children with behavior disorders. The schizophrenic children displayed the great majority of abnormal movements observed (toe walking, rocking, "clinging," and whirling).

575 **Lefcourt, Herbert M.; Rotenberg, Francine; Buckspan, Barbara; and Steffy, Richard A.** "Visual Interaction and Performance of Process and Reactive Schizophrenics as a Function of Examiner's Sex." *Journal of Personality* 35 (1967): 535-546. *T*

Thirty newly admitted male patients were classified as process or reactive schizophrenics on the basis of tests. The amount of eye contact they showed with a male and a female interviewer was observed. The reactive patients looked more frequently at the interviewers than did the process schizophrenics. The process patients looked at the woman less than at the man; yet they performed the Digit Span Subtest of the WAIS better with the female examiner than with the male.

576 **Levanthal, Howard, and Sharp, Elizabeth.** "Facial Expressions as Indicators of Distress." In *Affect, Cognition, and Personality,* edited by S. S. Tomkins and C. E. Izard, pp. 296-318. New York: Springer Publishing Co., 1965. *G N T*

The facial expressions of women during labor—nineteen for the first time ("primagravidae") and seventy-one who had been in labor before ("multigravidae") —were observed and recorded with a notation. The primagravidae mothers showed less comfortable expressions and more distress signs throughout. Also, the women tested and rated as high in anxiety showed more distress signs. Intercorrelations and reliability of the measures and medication effects are discussed.

577 **Leventhal, Marcia B.** "A Dance-Movement Experience as Therapy with Psychotic Children." Master's thesis, University of California, 1965.

The author evaluated the effects of individual dance therapy sessions with four hospitalized psychotic children over a twelve-week period. Figure drawings, hospital reports, and Hunt's Force, Time and Space Movement Scale were used for evaluation. Notwithstanding limitations of research design, improvement in body image and socialization was indicated.

578 **Levy, David M.** "Finger Sucking and Accessory Movements in Early Infancy: An Etiologic Study." *American Journal of Psychiatry* 7 (1928):881-918. *T*

An extensive study of finger sucking in infants, the forms it takes, what actions accompany it, and what early experiences are common to finger suckers. A survey of some 130 infants suggests that the sucking is due to insufficient lip activity while feeding. "Accessory movements" that accompany sucking are traced back to movement of hands at the breast. No relation was found between incidence of finger sucking and sex, social status, type of feeding, nutrition, or position in the family.

579 **Levy, David M.** "On the Problem of Movement Restraint: Tics, Stereotyped Movements, Hyperactivity." *American Journal of Orthopsychiatry* 14 (1944): 644-671. *T*

First, Levy reports observations that chickens restricted to cages showed more frequent and intense "head-shaking tics" than did free-ranging chickens. An experiment demonstrated that this was related to the amount of floor space. He then reviews literature on abnormal or stereotyped movements of horses, zoo animals, and institutionalized children as a function of the degree of physical restraint. He presents clinical records showing that periodic movement restraint may lead to hyperactivity; discusses clinical cases of "psychic" tic; and cites research on emotional responses to restraint.

580 **Levy, David M.** *Behavioral Analysis: Analysis of Clinical Observations of Behavior, As Applied to Mother-Newborn Relationships.* Springfield, Ill.: Charles C Thomas, Publisher, 1958. 370 pp. *T*

This must be the most comprehensive behavioral description and analysis of mother-infant interaction yet published. Observers made notes of all visible behavior, bodily expressions and sounds or verbalizations of infant and mother in two or three visits up to the child's eighth day. Much of these records is presented in chapters on the Initial Phase (greeting the baby), the Feeding Phase, Painful Breast Feedings, "Tension, Distraction, Passivity," and the End Phase, together with quantitative analysis of the recordings and various ratings on the nature of the relationship between mother and child, attitude of the mother and so on.

581 **Levy, Leon H.; Orr, Thomas B.; and Rosenzweig, Sanford.** "Judgment of Emotion from Facial Expressions by College Students, Mental Retardates, and Mental Hospital Patients." *Journal of Personality* 28 (1960):342-349. *T*

Although "median judgments" of the degree of happiness or unhappiness expressed in a number of photographs were highly correlated between groups of college students, high-functioning retardates, and acutely disturbed psychiatric patients, the judgments of the clinical groups were more heterogeneous.

582 **Lipsitt, Lewis P., and De Lucia, Clement A.** "An Apparatus for the Measurement of Specific Response and General Activity of the Human Neonate." *American Journal of Psychology* 73 (1960):630-632. *D P*

Describes a stabilimeter apparatus for measuring the amount of the infant's activity and an apparatus for measuring leg movements.

583 **Loeb, Felix F.** "The Psychoanalytic Significance of Grasp-Like Movements as Communicational Signals." Unpublished paper, Pittsburgh Psychoanalytic Institute, 1966. 36 pp. *D T* (Copies from author, University of Pittsburgh School of Medicine, Western Psychiatric Institute and Clinic, 3811 O'Hara St., Pittsburgh, Pa. 15213)

From three interview films, minute film analysis of a recurring grasping gesture and the words accompanying it indicated that it was associated with a specific lexical context and a "meaning class" of "in close to" or "away from." Observations of greeting patterns, praying movements, and movements of human and animal infants nursing, as well as grasplike movements of psychoanalytic patients, are cited as evidence that this movement is related to grasping movements at the breast or the wish to be close to the mother's breast. Loeb discusses the phylogenetic and psychoanalytic significance of his observations.

584   Loeb, Felix F. "Rooting-Like Movements in an Adult Human: A Frame-by-Frame Study." Unpublished paper, n.p., 1968. 27 pp. *T* (Copies from author, University of Pittsburgh School of Medicine, Western Psychiatric Institute and Clinic, 3811 O'Hara St., Pittsburgh, Pa. 15213)

Two different types of side-to-side head shakes of a patient in a psychotherapy session were carefully analyzed from film. The author describes a "yes-type" shake (more forward), which was found to be associated with verbal expressions of want, and a "no-type" one (no forward component), which was the conventional expression of negation. The observations are discussed within the framework of psychoanalytic assessment of the infant's rooting response and early development.

585   Loeb, Felix F. "The Microscopic Film Analysis of the Function of a Recurrent Behavioral Pattern in a Psychotherapeutic Session." *Journal of Nervous and Mental Disease* 147 (1968):605-617. *T*

Integrating psychoanalytic formulations and the context-analysis method of studying interactional behavior defined by Scheflen, the author reports on film study of a recurring movement, the fistlike gesture of a woman observed in a therapy session. Detailed systematic analysis showed that it occurred only with words explicitly or preconsciously denoting anger, frustration, or fear of harming her son.

586   Loeb, Felix F.; Loeb, Loretta R.; and Ross, David S. "Grasping as an Adult Communicational Signal." *Journal of Nervous and Mental Disease,* in press.

Subjects were shown films of the grasplike movement Loeb had previously correlated with the infant's hand movements while nursing and other arm movements similar to it and were asked to free-associate to the meaning of the different movements. The authors' analysis of the subjects' associations shows that the experimental gesture was significantly associated with categories labeled "on-off," "negation," "yes," and "finished." The authors discuss their findings in relation to Spitz's writings on infant development of expressions of negation.

587   Lomax, Alan; Bartenieff, Irmgard; and Paulay, Forrestine. "Dance Style and Culture." In *Folk Song Style and Culture* by A. Lomax, pp. 222-247. Washington, D.C.: American Association for the Advancement of Science, Publication no. 88, 1968. *G T*

The authors present the rationale for their research, using the choreometrics analysis of dance and everyday activity that they had developed. The primary hypothesis is that the dance style of a given culture is a "crystallization" of those patterns of work and activity which are most prevalent and important to the group. Observing anthropological films, they found the style of work identical to that of dance in most cases. Certain movement characteristics, such as the type of body attitude, distinguish eight world regions in a way similar to that marked by characteristics of song style. Further, the type of transition in movement, the number of body parts, and the level of complexity in terms of "effort" and "shape" qualities of movement correlate highly with the level of production or subsistence complexity of the culture.

588   Lomax, Alan; Bartenieff, Irmgard; and Paulay, Forrestine. "The Choreometric Coding Book." In *Folk Song Style and Culture* by A. Lomax, pp. 262-273. Washington, D.C.: American Assiciation for the Advancement of Science, Publication no. 88, 1968.

This chapter is a presentation of the *Choreometric Coding Book*, which includes definitions of each of the terms of the choreometrics system and instructions on how to observe movement with the Choreometric Coding Sheet. This coding sheet is a checklist of features or qualities of body movement used for analyzing dance and everyday activity. A method for studying features of movement which are shared by entire cultures or world regions, it is composed of terms for how the body is used, body attitude, the "shape" and the "energy" of the transition and main activity, and the "spread of flow through the body." A sample of the coding sheet is reprinted herein.

589   Long, John A. *Motor Abilities of Deaf Children*. New York: Bureau of Publications, Teachers College, Columbia University, 1932. 67 pp. *D T*

The motor abilities of deaf and hearing schoolchildren were compared, using the apparatus and simple motor tasks of the Stanford Motor Skills Unit, described here in detail. The deaf children were superior in the "tapping, motility rotor, and spool-packing" tasks and inferior on the balancing test and a serial discrimination perceptual-motor task.

590   Loraas, O. Ivar. "The Relationship of Induced Muscular Tension, Tension Level and Manifest Anxiety in Learning." *Journal of Experimental Psychology* 59 (1960):145-152. *G T*

In a learning experiment it was found that subjects rated highest on the Taylor Manifest Anxiety Scale had the highest blink rates (the measure here of muscular tension), and a number of relationships were found between high and low anxiety, muscular tension, induced muscular tension, and performance in paired-associate learning.

591   Lorenz, Konrad. *On Aggression*. Translated by M. K. Wilson. New York: Harcourt Brace Jovanovich, 1966. 306 pp. *D*

In this masterwork on aggression in man and animals, Lorenz distinguishes between aggression among species, such as when a hungry predator stalks its prey, and within a species, such as in fights for territory, sexual selection, or brood defense, and he points out that the aggressive movements are different in the two cases (pp. 25, 40, 45). The chapters on ritual are particularly relevant here, including the fascinating account of the "inciting" movements of ducks, how these evolve and become "automatic" rituals that are ends in themselves, and the evolution of human rituals and manners and their function in controlling aggression, forming group bonds, and distinguishing groups. Lorenz discusses the degree to which fixed motor patterns or "tool activities" have survival value apart from their relation to drives. In one part, he analyzes animal movements of sexuality and aggression in terms of their spatial directions. There is a wealth of descriptive examples of fight-flight movements, "courtship" behaviors, meeting behaviors, appeasement gestures, and modes of bonding and group formation among fish, ducks, deer, birds, lizards, rats, dogs, and other animals.

592   Lott, Dale F., and Sommer, Robert. "Seating Arrangements and Status." *Journal of Personality and Social Psychology* 7 (1967):90-95. *T*

One study in a series on this subject showed that subjects sat closer to a surrogate of someone they thought was a peer and further from either a high- or a low-status individual.

593   Lowen, Alexander. *The Language of the Body*. (Grune & Stratton, 1958; reprint ed., New York: The Macmillan Company, 1971). 400 pp. *D*

Originally published in 1958 under the title *Physical Dynamics of Character Structure*, this is one of the most advanced studies of movement patterns, muscular tension, posture, and body physique in relation to personality. A former student of Wilhelm Reich, Lowen greatly elaborates and extends his writing on body expression and character, and presents theoretical chapters integrating analysis of body movement and muscular tension patterns with psychoanalytic psychology. The theory and practice of what Lowen calls "bioenergetic analysis" is illustrated with numerous case examples. Postures and movement characteristics of specific character types (oral, masochistic, hysterical, phallic-narcissistic, passive-feminine, schizophrenic, and schizoid) are analyzed in relation to the patient's history and psychodynamics. This book contains a wealth of interpretations and hypotheses about the significance of specific muscular tensions.

594 **Lowen, Alexander.** *Love and Orgasm.* (The Macmillan Company, 1965; reprint ed., New York: New American Library, 1967). 319 pp. *D*

Lowen expands Reich's theories of the relation between mature heterosexuality and mental health and maturity, and that of various forms of homosexuality, frigidity, and impotence to neuroses and severe intrapsychic conflicts. Complete, spontaneous orgasm becomes the reflection of a person who is integrated, loving, unafraid of closeness. Throughout this treatise on love and sexuality, Lowen underlines the importance of the body and movement in reflecting personality and sexual development. Cases of individuals with various sexual problems are described, including analyses of their physiques and characteristic postures and expressions. There is also an analysis of sexual intercourse, with a description of the transition from "mobility" (consciously directed movement) to "motility" (immediate, involuntary expression of feeling in movement).

595 **Lowen, Alexander.** *Betrayal of the Body.* New York: The Macmillan Company, 1967. 307 pp. *D*

The major work on the relation between schizoid conditions and disturbances of motor activity, breathing, and body image. With numerous case descriptions, Lowen illustrates how he assesses the expressive behavior, patterns of muscle tension, breathing, and body structure of his patients and correlates them with the patients' figure drawings, early history, psychological symptoms, and their introspective reports. He makes formulations as to the origins and defensive function of the schizoid physical behavior and analyzes the relationship between these physical characteristics and experiences of unreality and illusion. Movement behavior of schizophrenic patients is also noted. "Bio-Energetic" techniques for treating the physical behavior directly are described in a chapter entitled "Reclaiming the Body."

596 **Lourie, Reginald S.** "The Role of Rhythmic Patterns in Childhood." *American Journal of Psychiatry* 105 (1949):653-660.

Drawing on observations of 130 children, the author describes a variety of repetitive, rhythmic movements that may appear in normal childhood (e.g., "sleep rocking," finger tapping, head rolling). He discusses their relationship to developmental processes, tension reduction, frustration, restraint, breathing and heart rates, and special physical and psychiatric problems. He also reviews several theories concerning the origins and function of rhythmic movement in development.

597 **Luborsky, Lester; Blinder, Barton; and Schimek, Jean.** "Looking, Recalling, and GSR as a Function of Defense." *Journal of Abnormal Psychology* 70 (1965): 270-280. *D T*

With the Mackworth eye camera, which films the picture and the subject's points of visual fixation on it, records were obtained of the looking patterns of sixteen subjects. Those previously tested as having the defense of isolation looked around more; those with the defense of repression restricted their visual scanning and focused mainly on the figure alone. Also, "the defense of isolation was better than repression for recall of sexual content." (p. 279)

598   Lundervold, Arne. "An Electromyographic Investigation of Tense and Relaxed Subjects." *Journal of Nervous and Mental Disease* 115 (1952):512-525. *D G P*

EMG recordings of different muscle groups were made of "tense" and "relaxed" subjects, 110 Norwegian men and women while .typing. The apparatus and EMG records are well illustrated, but there is limited analysis of psychological factors.

599   Luria, Alexander R. *The Nature of Human Conflicts: An Objective Study of Disorganization and Control of Human Behaviour.* Translated and edited by W. H. Gantt. New York: Liveright Publishing Corporation, 1932. 431 pp. *D G P T*

The study of affect and behavior organization from "active forms of human activity," specifically voluntary movements which are considered reflections of concealed psychological processes, is the subject of this research. Luria describes an apparatus in which the subject rests his right hand and fingers on a pneumatic bulb, which he presses when given a stimulus or asked to associate to words, while his left hand stays passive on a weight. Subtle changes in the tremor of both hands are thus recorded on a kymograph. A great deal of research is reported here, notably how the motor patterns are used to study lie detection and affect in criminal suspects, thought patterns and inhibited associations, affective processes under trauma and stress, personality characteristics, developmental stages, and psychodiagnosis. The subjects are students waiting to hear if they have failed, prisoners recently arrested for major crimes, hypnotized subjects who are given disturbing suggestions, and young children given rhythm or simple motor tasks to perform. These experiments are often disturbing to read about. The discussion about the relation between movement, affect, and thought processes is brilliant.

600   Luria, Aleksandr R. *Higher Cortical Functions in Man.* Translated by B. Haigh. New York: Basic Books, 1966. 513 pp. *D P T*

In this monumental work, Luria analyzes the effects of local brain lesions on higher mental functions, including disturbances of body movement and kinesthetic processes. Notwithstanding the scope and difficulty of the subject, this book is made readable and fascinating by the author's style, his focus on clinical rather than experimental evidence, and the extraordinary section on "bedside" tests he utilizes to assess specific lesions, including a number of ingenious structured motor tests for laterality, coordination, strength, spatial organization of movement, complex patterns, etc. Since the neurophysiology of movement is not specifically included in this present bibliography, the reader is referred to Luria's review of the literature and his extensive bibliography.

601   Luthe, Wolfgang, and Schultz, Johannes H. *Autogenic Therapy,* Vol. 3, *Applications to Psychotherapy.* New York: Grune & Stratton, 1969. 228 pp. *D G P T*

The effects of autogenic relaxation and concentration methods in the treatment of a wide range of neurological and psychiatric disorders are presented.

602 **Lynn, John G.** "An Apparatus and Method for Stimulating, Recording and Measuring Facial Expression." *Journal of Experimental Psychology* 27 (1940):81-88. *D P*

The apparatus described and illustrated combines a film projector with earphones and a high-speed camera. The subject looks through an opening at a screen; the film he sees varies according to what expressions are to be stimulated; and a camera films his expressions without his being aware. A reliable method for measuring details of facial expressions (mouth, eyes, forehead, etc.) from film frames projected on a screen is described.

603 **Lynn, John G., and Lynn, Doris R.** "Smile and Hand Dominance in Relation to Basic Modes of Adaptation." *Journal of Abnormal and Social Psychology* 38 (1943):250-276. *D G T*

In an initial study, subjects (children and particularly adults) who showed a "homolateral" smile-hand dominance pattern (e.g., "right-smiledness" with right-handedness) versus those with a "contralateral" pattern showed a specific pattern of reaction, including greater self-confidence, initiative, independence, etc., as compared with the "contralateral" group. A facial cinerecorder apparatus and methods of objectively studying smile laterality are described. Similar results were found in a group of hospitalized psychiatric patients, along with evidence that the degree of "smiledness" in general correlates with the degree of these personality traits.

604 **Lynn, R.** *Attention, Arousal, and the Orientation Reaction.* Oxford, Eng.: Pergamon Press, 1966. 118 pp. *D G T*

Experimental research on the orientation reaction of animals and humans to novel, threatening, or interesting stimuli. Its visceral and motor aspects are described, and underlying physiological mechanisms are assessed. Chapters on individual differences, developmental aspects, and the role of the orienting response in conditioning are included.

605 **Machotka, Pavel.** "Body Movement as Communication." In *Dialogues: Behavioral Science Research*, vol. 2. Boulder, Colo.: Western Interstate Commission for Higher Education, 1965. pp. 33-66. *D N*

Subjects' interpretations of the body positions of figures in drawings and classical paintings are reported. The author first presents definitions and notations for various positions and distances between figures. Noting the factors that appear to influence the judges' interpretations, he describes how differently subjects judge a female figure in which arm positions are varied and how direction of gaze, proximity, and position affect judgments of social relationships between figures.

606 **Maginnis, Maria.** "Gesture and Status." *Group Psychotherapy* 11 (1958):105-109.

A summary of a very interesting Ph.D. dissertation on the number and kinds of gestures observed in schoolchildren, as compared with their rank from sociometric ratings by their peers. In addition to tabulating the number of different, identical, "random," and "common" gestures the children exhibited in class discussions, thirteen types of gestures were identified (e.g., "startle residue," "irrational," "expansive"). The "underchosen" child proved to have fewer "common" gestures and more intrapunitive actions (biting, pulling at his body, etc.), and the underchosen boy of low intelligence showed more "startle residue" behaviors (stares, fist clenching, etc.). Movements characteristic of the "overchosen" tended to be those common to the group, but these varied by sex and age. For example, the older overchosen intelligent boy tends to have "expansive" gestures (raises eyebrows, stretches arms behind chair), and the older overchosen girl tends to have more "masticatory" and "extrapunitive" gestures.

607 **Magriel, Paul D.,** compiler. *A Bibliography of Dancing.* (H. W. Wilson Company, 1936; reprint ed., New York: Benjamin Blom, 1966). 229 pp. *D*

This classic bibliography of dance groups the titles, many of which are annotated, according to eight major areas (e.g., History and Criticism, Ethnological, Mime and Pantomime) and has an extensive author and subject index. It is a good early source for references on ethnic dance, which are listed by country and group. A supplement cover the years 1936-1940.

608 **Mahl, George F.** "Gestures and Body Movements in Interviews." In *Research in Psychotherapy,* vol. 3, edited by J. M. Shlien. pp. 295-346. Washington D.C.: American Psychological Association, 1968. *D T*

Three exploratory studies are reported here, involving observation of gestures and postural shifts in psychiatric interviews or psychoanalytic sessions. The first involved interviews and the author's making a running account of the patient's movements observed without sound. General postural changes, "communicative" gestures, and "autistic" actions were then interpreted by the author, and predictions concerning psychodiagnosis, areas of conflict, emotional states, etc., were made. Tables presented show (1) marked male-female differences in the occurrence of certain gestures and positions; (2) the movements, their interpretations, and what in some cases are dramatic corroborations with clinical records; and (3) the appearance and frequency of certain acts of individual patients. The author summarizes this pilot study of eighteen patients with some hypotheses regarding individual and sex differences in nonverbal behavior and the relationship between speech and motion. In a second section, he presents a schematic analysis of certain "autistic" actions observed in psychoanalytic sessions and their relation to unconscious fantasies, subsequent verbalizations, early childhood conflicts, dreams, and defenses. Third, he reports on an experiment in which students were interviewed sitting face-to-face and then back-to-back, once with a female interviewer and once with a male. In the back-to-back situation, there was an increase in "autistic" gestures, although most subjects did not feel more uncomfortable; the amount and expansiveness of postural shifts was the same; and in some cases, the movement was less inhibited. Mahl concludes with a case history in which the personal, psychodynamic meanings of a symbolic hand gesture are elucidated.

609 **Mahl, George F.; Danet, Burton; and Norton, Nea.** "Reflection of Major Personality Characteristics in Gestures and Body Movement." *American Psychologist* 14 (1959):357.

This is a very brief summary of a conference paper on the relation of body movements to personality and sex differences, as well as to the "meaning" of the verbal content in psychiatric interviews.

610 **Mahler, Margaret S.** "Tics and Impulsions in Children: A Study of Motility." *Psychoanalytic Quarterly* 13 (1944):430-444.

Following a psychoanalytic discussion of the role of motility in early development (with references to a number of German articles not cited here), the author defines and discusses "impulsions," or forms of "immediate acting out," and repetitive actions and tics. The psychodynamics and origins of tics are illustrated with case examples. The early body movement characteristics and pre-tic behaviors of these children, notably poor motor coordination and hypermotility, are described; three types of tic patients are delineated according to when the sympton appears.

611   Mahler, Margaret S.; Luke, Jean A.; and Daltroff, Wilburta. "Clinical and Follow-up Study of the Tic Syndrome in Children." *American Journal of Orthopsychiatry* 15 (1945):631-647.

The classical tic syndrome ("involuntary, lightning-like repetitious jerks" of a muscle group) was studied clinically in sixteen children, and then in eighteen children of a follow-up investigation. Family characteristics, sex and religious differences, characteristic developmental problems, and movement and activity patterns are reported in detail. The brief case histories and discussion provide considerable information about the etiology, variations, and prognosis of the syndrome. The bibliography cites a number of English-language and German articles on the tic syndrome.

612   Maisel, Edward (selected and introduced by). *The Resurrection of the Body: The Writings of F. Matthias Alexander.* New Hyde Park, N.Y.: University Books, 1970. 204 pp. *D*

Alexander addressed himself to the analysis of "one's manner of use" and psychophysical patterns (carriage, areas of body constriction, etc.), as these reflect and affect emotion and thinking. Within the sections from his four books reprinted here, there are chapters on how he treated these habits in what has become known as the "Alexander technique," including the case history and treatment of a man who stuttered. Included also is a chapter by Maisel on Alexander, his theories and technique.

613   Mallery, Garrick. *Introduction to the Sign Language among the North American Indians as Illustrating the Gesture Speech of Mankind.* Smithsonian Institution Bureau of Ethnology. Washington, D.C.: Government Printing Office, 1880. 72 pp.

Includes a description of the gestures and what they stand for; how a story is told by using them; discussion of differences among tribes; and syntactical aspects of the sign languages.

614   Mallery, Garrick. "The Gesture Speech of Man." *Proceedings of the American Association for the Advancement of Science* 30 (1881):283-313.

In this address on gesture languages, particularly American Indian sign language, Mallery refutes then-current myths about the origins and usage of Indian sign language and makes comparisons of it with deaf sign language. He refers to a number of very old anthropological reports of gestures, to some literary and dramatic examples, and to the relation between speech and gesture.

615   Malmo, Robert B. "Experimental Studies of Mental Patients under Stress." In *Feelings and Emotions*, edited by M. L. Reymert, pp. 169-180. New York: McGraw-Hill Book Company, 1950. *G P T* [New York: Hafner Publishing Company]

This chapter summarizes three earlier investigations of finger and head movements and EMG, EKG, and respiratory patterns in normal subjects, early schizophrenic groups, psychiatric patients with marked anxiety and those with less evidence of anxiety, and patients with somatic complaints. Movement and muscle-potential disturbances and breathing irregularity correlated with anxiety; heart rate and also breathing and heart rate variability correlated with cardiovascular complaints.

616   **Malmo, Robert B.; Boag, Thomas J.; and Raginsky, Bernard B.** "Electromyographic Study of Hypnotic Deafness." *Journal of Clinical and Experimental Hypnosis* 2 (1954):305-317. *G T*

A comparison of EMG recordings of reactions to loud sound for subjects with hypnotically induced deafness, a patient with hysterical deafness, and a genuinely organically deaf person.

617   **Malmo, Robert B.; Boag, Thomas J.; and Smith, A. Arthur.** "Physiological Study of Personal Interaction." *Psychosomatic Medicine* 19 (1957):105-119. *G T*

EMG recordings were made of both patient and examiner during a TAT session and of patient and interviewer during a follow-up psychiatric interview. Half of the patients experienced criticism from the examiner, and half were praised. Speech muscle tension patterns indicated a greater "person-to-person uniformity in tensional reaction to criticism" (p. 117). Speech muscle tension also was found to correlate with frontalis tension and motor irregularities, but no general muscle tension factor was found.

618   **Malmo, Robert B.; Shagass, Charles; Bélanger, David J.; and Smith, A. Arthur.** "Motor Control in Psychiatric Patients under Experimental Stress." *Journal of Abnormal and Social Psychology* 46 (1951):539-547. *G T*

Evidence is presented supporting the hypothesis that "psychoneurotic disorders involve defective motor regulation manifested by abnormally increased motor disturbance under any stressful condition" (p. 540). A Luria-type method was used to obtain tremor pattern during rapid discrimination tests of psychoneurotics, acutely psychotic patients, chronic schizophrenics, and normal controls. Pressure, synchrony between right and left finger movements, and irregularity of pattern were analyzed.

619   **Malmo, Robert B.; Shagass, Charles; and Davis, Frederick H.** "Symptom Specificity and Bodily Reactions during Psychiatric Interview." *Psychosomatic Medicine* 12 (1950):362-376. *G T*

EMG recordings of muscle tension patterns in three psychiatric patients during their psychiatric interviews showed a clear relationship between increase in muscle tension and psychological stress and specific reaction in body parts associated with somatic complaints. Variations in tension patterns within and between interviews are illustrated in detail.

620   **Malmo, Robert B.; Shagass, Charles; and Davis, John F.** "Electromyographic Studies of Muscular Tension in Psychiatric Patients under Stress." *Journal of Clinical and Experimental Psychopathology* 12 (1951):45-66. *G T*

EMG recordings of muscle tension of chronic schizophrenics, acute psychotics, psychoneurotic patients, and normal controls were made during stress tests (pain, rapid discrimination, and mirror drawing). Muscle tension was higher for the neurotic and psychotic patients as compared with controls. Evidence that the muscle tension measures are independent of Luria tremorgram measures is presented. Individual consistency over several stress situations was also found.

621   **Malmo, Robert B., and Smith, A. Arthur.** "Forehead Tension and Motor Irregularities in Psychoneurotic Patients under Stress." *Journal of Personality* 23 (1955):391-406. *T*

EMG recordings were compared with finger tremor measures, observations of head and body movement, breathing irregularities, and heart rate during reaction to painful stimuli. The subjects were headache-prone and nonheadache-prone psychiatric patients and normal controls. Frontalis-muscle tension and "movement irregularities" distinguished patients from controls. Factor analysis of the data supported other findings of no correlation between muscle tension and other movement measures and also showed no correlation between frontalis tension and tension in other body areas.

622 **Malmo, Robert B.; Smith, A. Arthur; and Kohlmeyer, Werner A.** "Motor Manifestation of Conflict in Interview: A Case Study." *Journal of Abnormal and Social Psychology* 52 (1956):268-271. *T*

Further evidence that forearm muscle tension is related to hostility topics and that leg tension is associated with sex content.

623 **Malmo, Robert B.; Wallerstein, Harvey; and Shagass, Charles.** "Headache Proneness and Mechanisms of Motor Conflict in Psychiatric Patients." *Journal of Personality* 22 (1953):163-187. *G T*

EMG recordings during pain stimulation showed psychiatric patients with complaints of head and neck pain showed more evidence of "central conflict" than did nonheadache-prone patients. Also, when pain was induced at the forehead, only the headache-prone group displayed a shift of tension to the neck muscles.

624 **Marañon, Gregorio.** "The Psychology of Gesture." *Journal of Nervous and Mental Disease* 112 (1950):469-497.

First written in Spanish before World War II, this article deals with mass "contagion" of emotion through movement, the importance of gestures in conveying emotion and in arousing and organizing crowds, the dictator's use of gesture, and the "collective gestures" and rhythms that reflect loss of individuality and organization of the masses. The relation of collective gestures and rhythms to military discipline and to hypnosis is discussed, along with prerequisite conditions for the power of gestures to influence the masses.

625 **Marler, Peter R.** "The Logical Analysis of Animal Communication." *Journal of Theoretical Biology* 1 (1961):295-317.

Animal facial expressions and body signals discussed in the light of communication theory.

626 **Marler, Peter.** "Communication in Monkeys and Apes." In *Primate Behavior: Field Studies of Monkeys and Apes*, edited by I. De Vore, pp. 544-584. New York: Holt, Rinehart and Winston, 1965. *D G T*

Much of the information in this chapter on how primates interact and signal each other deals with touch, facial displays, body movements, and postures, and is well-illustrated with photographs and drawings. The function of these behaviors in dominance patterns, mating behavior, and group regulation is discussed.

627 **Marler, Peter.** "Animal Communication Signals." *Science* 157 (1967): 769-774.

An analysis of the communicative function of animal signals in terms of spatial orientation to the signal, behavior elicited in the recipient, contexts in which the responses recur, and the relationship between the signal and response.

628 **Marquis, Dorothy Postle.** "A Study of Activity and Postures in Infants' Sleep." *Journal of Genetic Psychology* 42 (1933):51-69.

An analysis of the amount of activity and types of postures and movements of thirteen infants during day and night sleep, using measuring apparatus and observation.

629   **Martin, Donald W., and Beaver, Nohmie**. "A Preliminary Report on the Use of Dance as an Adjuvant in the Therapy of Schizophrenics." *Psychoanalytic Quarterly Supplement* 21 (1951):176-190.

This description of dance sessions involving relaxation techniques and social dance instruction with a group of young schizophrenic outpatients demonstrates their individual reactions and the improvement in eight of the patients.

630   **Martin, Irene**. "Personality and Muscle Activity." *Canadian Journal of Psychology* 12 (1958):23-30. *T*

Four groups were selected who had extreme scores on questionnaires for rating neuroticism and introversion, and EMG recordings of forehead and forearm were taken while they were at rest and while they answered questions about neurotic symptoms. No significant differences in muscle tension were discerned among the groups (introverted neurotics, extroverted neurotics, introverted normals, extroverted normals), although greater tension was found while they were talking even when there was no overt movement.

631   **Martin, John**. *Introduction to the Dance*. (W.W. Norton & Company, 1939; reprint ed., New York: Dance Horizons, 1965). 363 pp. *P*

Discussion of dance aesthetics, historical and cultural origins and various styles, etc., but unfortunately there is little explicit description of dance and movement.

632   **Maslow, A. H**. "The Expressive Component in Behavior." *Psychological Review* 56 (1949):261-272.

A strong distinction is made here between "coping" (defined as instrumental, purposive) and "expressive" (noninstrumental, unmotivated, unlearned) behavior. Expressive behavior—exemplified here by numerous references to postures, movements, and facial expressions, among other aspects of behavior—is defined as an "epiphenomenon" of character that "mirrors, reflects, signifies, or expresses some state of the organism." Further distinctions are drawn between expressive behavior that releases tension versus coping responses motivated by basic needs. A relationship is posited between rituals, tics, symbolic acts, etc., and repetitious attempts to complete coping acts. The implications of this expressive-coping distinction for psychodiagnosis and treatment are discussed.

633   **Mason, William A**. "The Social Development of Monkeys and Apes." In *Primate Behavior: Field Studies of Monkeys and Apes,* edited by I. De Vore, pp. 514-543. New York: Holt, Rinehart and Winston, 1965. *D G*

From a review of primate literature and recent laboratory experiments, the author surveys the development of primate social behavior: stages of infant social responses and mother-infant interaction, forms of play, effects of deprivation, the significance of repetitive movements and stereotyped behaviors, and development of sexual behavior.

634   **Matarazzo, Joseph D.; Saslow, George; Wiens, Arthur N.; Weitman, Morris; and Allen, Bernadene V**. "Interviewer Head Nodding and Interviewee Speech Durations." *Psychotherapy: Theory, Research and Practice* 1 (1964):54-63. *T*

In two groups of twenty subjects having employment interviews, the interviewers nodded their heads all the while the subject spoke during the second fifteen-minute interval. In a third control group, the interviewer did not do this. The head nodding resulted in a 48 to 67 percent increase in the duration of a subject's single utterances, whereas there was no such increase in the control group.

635 **Mawer, Irene.** *The Art of Mime: Its History and Technique in Education and the Theatre.* London: Methuen & Co., 1932. 244 pp. *D P*

Includes chapters on the origins and history of mime in Asiatic, Egyptian, Greek, and Roman drama and in European theater over several centuries, along with sections on technique and interpretation of mime gestures, foot and leg movements, and head and facial expression.

636 **May, Philip R. A.; Wexler, Milton; Salkin, Jeri; and Schoop, Trudi.** "Non-Verbal Techniques in the Re-Establishment of Body Image and Self Identity: A Preliminary Report." *Psychiatric Research Reports* 16 (1963):68-82. *T*

A mode of therapy using movement, called the "body ego technique," is described and its effectiveness evaluated. A typical session using this technique to guide the patients through movement patterns of normal development, expressions, and attitudes and to support their body ego boundaries is described and contrasted with "expressive dance therapy." A research project in which twenty very disturbed schizophrenic patients participated in individual or group movement sessions and a control group received music therapy, as reported here, indicates improvement over six months in the movement group.

637 **McCaskill, Carra L., and Wellman, Beth L.** "A Study of Common Motor Achievements at the Preschool Age." *Child Development* 9 (1938):141-150. *T*

Ninety-eight children from two to six years old were given a series of structured tasks such as hopping, jumping, and ball throwing, to study developmental levels of motor skill, or "motor age." Age and sex differences are noted, and a tentative 73-stage list of skills with age assignments is presented.

638 **McCraw, Charles B.** *Scoreography.* Ann Arbor, Mich.: Edward Brothers (Lithographers), 1964. 58 pp. *D N*

A movement notation system based on music notation, with three shapes of "notes" indicating direction, symbols for body parts and touch, and placement on the staff indicative of the area of the body focused on.

639 **McGinnis, Esther.** "The Acquisition and Interference of Motor Habits in Young Children." *Genetic Psychology Monographs* 6 (1929):203-311. *D G T*

A study of motor learning in children ages three to five, using stylus mazes. It is particularly notable here for its extensive bibliography and review of the literature on motor ability and learning. A table of children's motor studies from 1872 to 1928 furnishes information on populations used, procedures, and findings; and another table lists psychomotor tests from 1911 to 1927.

640 **McGinnis, John M.** "Eye-Movements and Optic Nystagmus in Early Infancy." *Genetic Psychology Monographs* 8 (1930):321-430. *G P T*

This monograph begins with an excellent review of literature on infant eye movements and methods for recording eye movements. The apparatus used in this film study of six neonates from birth to forty-two days is elaborately described. Patterns of eye and head movement and their development are reported—notably when, how, and under what conditions optic nystagmus and ocular pursuit appear.

641 **McGraw, Myrtle B.** *Growth: A Study of Johnny and Jimmy.* New York: Appleton-Century-Crofts, 1935. 319 pp. *D G P*

This classic study of the development of infant twins contains elaborate analysis of the motor patterns of the twins, the effects of training at different periods, and developmental stages of motor functioning: reflexes, prehension, locomotion. Richly illustrated with photographs and charts.

642 McGraw, Myrtle B. *The Neuromuscular Maturation of the Human Infant.* New York: Columbia University Press, 1943. 140 pp. *D G* [New York: Hafner Publishing Company]

An important study of general developmental changes in posture, prehension, and locomotion in the infant and young child, with formulations about parallels ·with neurophysiological and anatomical developments. It includes analysis of developments from various reflexes into deliberate motor control, particularly of grasping behavior, "aquatic behavior," postural adjustments, and locomotion. Normal development of uprightness and locomotion is elaborately documented and illustrated. Chapters on visual-motor development and individual differences are included.

643 McKinley, J. Charnley, and Berkwitz, N. Joseph. "Quantitative Studies on Human Muscle Tonus: I, Description of Methods." *A.M.A. Archives of Neurology and Psychiatry* 19 (1928):1036-1056. *D G P*

A complicated apparatus to measure tonus while the arm is passively pulled, methods of analyzing the kymograph data, and pattern variations in normal and neurologically impaired subjects are described and illustrated.

644 Mead, George H. *Mind, Self, and Society.* Edited by C. W. Morris. Chicago: University of Chicago Press, 1934. 401 pp.

In a discussion entitled "Wundt and the Concept of Gesture" (pp. 42-61), Mead analyzes the gesture as "that part of the social act which serves as a stimulus to other forms involved in the same social act" (p. 42). Using examples of non-verbal encounters between animals or humans, in an essay that becomes a discussion of the nature of communication and thinking, the author clarifies the relation between the "expressive" and "communicative" aspects of gestures, body movements, and facial expressions.

645 Mead, Margaret. *Coming of Age in Samoa.* New York: William Morrow & Company, 1928. 297 pp. *P* [New York: Dell Publishing Co., 1967]

Includes a chapter on the role of dance in the education of Samoan children.

646 Mead, Margaret. "From Intuition to Analysis in Communication Research." *Semiotica* 1 (1969):13-25.

A discussion of changes in research approaches to cultural style, including some recent anthropological studies of movement and dance.

647 Mead, Margaret, and Byers, Paul. *The Small Conference: An Innovation in Communication.* The Hague: Mouton Publishers, 1968. 126 pp. *P*

A most interesting book on the evolution of the small conference as a new form of communication, it has a photographic essay combining detailed analysis of conference interaction and nonverbal behavior with photographs of three different conferences. Few books are this successful in illustrating clearly what the kinesicist attends to.

648 Meerloo, Joost A. M. *Dance Craze and Sacred Dance.* London: Peter Owen, 1961. 151 pp. *D P* [New York: Humanities Press, 1962]

Essentially a sumptuous collection of photographs and paintings of people and objects moving—dancers around the world, mime artists, children, crowds, etc.—with the author's comments and interpretations.

649 **Meerloo, Joost A. M.** "Rhythm in Babies and Adults: Its Implications for Mental Contagion." *A.M.A. Archives of General Psychiatry* 5 (1961):169-175.

The importance of rhythms in normal development focusing on rhythmic floating in utero, rhythmic movements between mother and infant, biological rhythms, group rhythms in marching, dancing, and talking. A discussion of "mental contamination," as "induction of the same rhythm" and examples of automatic imitation of movements as a reaction to panic or catastrophy conclude the article.

650 **Meerloo, Joost A. M.** *Unobtrusive Communication: Essays in Psycholinguistics.* Assen, The Netherlands: Van Gorcum, 1964. 198 pp.

Contains sections on the role of rhythmic behavior and dance in "group contagion."

651 **Mehrabian, Albert.** "Orientation Behaviors and Nonverbal Attitude Communication." *Journal of Communication* 17 (1967):324-332. *T*

Pairs of subjects were interviewed by an experimenter who either maintained a head and body orientation toward one of them or split it (head toward one, body toward the other). Their judgments of which one the experimenter had a positive attitude toward showed that head-orientation "immediacy" is a significant factor in such judgments.

652 **Mehrabian, Albert.** "Relationship of Attitude to Seated Posture, Orientation and Distance." *Journal of Personality and Social Psychology* 10 (1968):26-30. *T*

Subjects were asked to assume postures they considered typical for them in relation to someone intensely liked, moderately liked, etc., ranging to intensely disliked. Eye contact, distance, subjects' head, shoulder, and leg orientation, the openness of their arms and legs, and the relaxation of hands, legs, and trunk (i.e., tilted) were rated by three experimenters. Variations in eye contact, distance, orientation of shoulders (asymmetrical or "head on"), and uprightness versus body leaning significantly indicated liking, whereas the other variables were not good indicators of liking or disliking.

653 **Mehrabian, Albert.** "Inference of Attitudes from the Posture, Orientation, and Distance of a Communicator." *Journal of Consulting and Clinical Psychology* 32 (1968):296-308. *T*

In the first experiments, given photographs of people of differing age, sex, and status seen in various positions and degrees of relaxation, subjects judged what the attitude of the photographed person toward themselves might be. In later experiments, subjects imitated ways they might stand in relation to imaginary persons differing in age, sex, status, and likableness. The nonverbal cues reflecting various status and attitude relationships are reported, and scoring criteria of standing postures are listed.

654 **Mehrabian, Albert.** "Methods and Designs: Some Referents and Measures of Nonverbal Behavior." *Behavior Research Methods and Instrumentation* 1 (1969):203-207. *T*

In a concise summary of research findings, the author describes which nonverbal cues (i.e., patterns of body position, posture, facial expression, distance, gaze behavior, touching, hand movements, and paralinguistic variations) indicate status, attitude toward addressee, relaxation, and responsiveness. He notes reliability figures and scoring criteria for each movement category.

655   **Mehrabian, Albert.** "Significance of Posture and Position in the Communication of Attitude and Status Relationships." *Psychological Bulletin* 71 (1969):359-372.

The author reviews some nonverbal research to glean from it what aspects of eye contact, posture, position and spacing reflect attitudes (positive or negative) and status. Positive attitude may be reflected in closeness, leaning forward, indirect body orientation (if male), a "medium" amount of eye contact, a moderate sideways lean, etc.; negative attitudes may be reflected in an "arms-akimbo" position, minimal eye contact, broad reclining position, etc. High status of the addressee appears to correspond with direct shoulder orientation, greater distance, and moderate eye contact of the communicator; and low addressee status, with indexes of relaxation, sideways lean, and minimum eye contact of the communicator.

656   **Mehrabian, Albert, and Ferris, Susan R.** "Inference of Attitudes from Nonverbal Communication in Two Channels." *Journal of Consulting Psychology* 31 (1967):248-252. *T*

Subjects were given a series of photographs and tape-recorded words, with the face and the tone of voice conveying positive, neutral, or negative attitudes. The facial component had a stronger effect on the subject's judgments than the tone of voice, but both were significant. The inferred judgments appear to be a weighted sum of the two channels.

657   **Mehrabian, Albert, and Friar, John T.** "Encoding of Attitude by a Seated Communicator Via Posture and Position Cues." *Journal of Consulting and Clinical Psychology* 33 (1969):330-336.

Subjects were asked to assume postures they would assume while interacting with a man or woman, liked or disliked and high or low in status (i.e., eight possible addressees). Results show that specific patterns of distance, body position, posture, and eye contact reflect positive attitude, whereas others depend on the status of the addressee. Sex differences are noted in the behavior of the communicators. A relationship between the degree of relaxation and the status of the addressee is described.

658   **Mehrabian, Albert, and Williams, Martin.** "Nonverbal Concomitants of Perceived and Intended Persuasiveness." *Journal of Personality and Social Psychology* 13 (1969):37-58. *T*

After three elaborate experiments which involved attempts of students to inform or persuade their peers are described, an analysis of their behavior is made from videotape. A number of nonverbal cues were found to reflect whether the speaker intended to be persuasive and was perceived as being persuasive. Factor analysis of the nonverbal behaviors and personality measures yielded such factors as Perceived Persuasiveness, Dominance, Relaxation, and Nonimmediacy. A number of correlations between specific nonverbal cues and psychological measures of neuroticism, dominance, comfort of the communicator, etc., are reported. Sex differences are noted in the results.

659   **Meige, Henry, and Feindel, E.** *Tics and Their Treatment.* Translated and edited by S. A. K. Wilson. London: Sidney Appleton, 1907. 386 pp.

A very thorough study of the psychological and physiological origins, nature, diagnosis, and definition of tics. The personality characteristics of tiquers are discussed, and numerous cases are described, including a long personal account of the illness by a tic patient. Various forms of tics and the development of specific tics are assessed. The book concludes with chapters on various modes of treatment of this motor disorder.

660 **Melcher, Ruth Taylor.** "Children's Motor Learning with and without Vision." *Child Development* 5 (1934):315-350. *D G P T*

In this controlled study of young children's ability to learn a maze visually or by manual guidance without vision, the manual method was found much less effective.

661 **Meltzoff, Julian; Singer, Jerome L.; and Korchin, Sheldon J.** "Motor Inhibition and Rorschach Movement Responses: A Test of the Sensory-Tonic Theory." *Journal of Personality* 21 (1953):400-410. *T*

Three experiments are reported in which the research subjects performed one of the Downey Will-Temperament Tests (writing as slowly as possible) between responses to Rorschach cards. The experimental groups showed a significant increase in M responses to the Rorschach after the motor-inhibition task.

662 **Metheny, Eleanor.** *Movement and Meaning.* New York: McGraw-Hill Book Company, 1968. 126 pp.

Theoretical exploration of movement as "the functional link between the subjective and objective components of human understanding" (p. 22) and the origins and sources of meaning in dance, sport, and exercise. The nature and interpretation of dance, the basis of sport forms in effective action defined by rules, and the function of exercise in effecting the condition of the performer are discussed.

663 **Michotte, Albert E.** "The Emotions as Functional Connections." In *Feelings and Emotions*, edited by M. L. Reymert, pp. 114-126. New York: McGraw-Hill Book Company, 1950. [New York: Hafner Publishing Company]

Michotte extends the results of research that originally investigated the perception of causation to an exploration of the nature of emotion. The experiments involved creating films of abstract figures moving about. Michotte found that, depending on the direction, distance, speed, and sequence of the moving rectangles, people would interpret them psychologically: e.g., A was frightened by B's approach and ran off. He lists the subtle differences in interpretation of affect and interaction created by only slight variations in the action sequence. He suggests that motor reactions of men or animals considered in relation to each other and to the environment are a crucial element in the perception and experience of emotion.

664 **Miles, W. R.** "Correlation of Reaction and Coördination Speed with Age in Adults." *American Journal of Psychology* 43 (1941):377-391. *D P T*

An example of a reaction-time study, together with a brief review of literature and description of apparatus. Differences in the speed of simple finger and foot movements were found in subjects from twenty-five to eighty-seven years old; but individual differences were also noted, and some of the oldest subjects performed as quickly as younger ones.

665 **Miller, Robert E.; Banks, James H.; and Kuwahara, Hiroshi.** "The Communication of Affects in Monkeys: Cooperative Reward Conditioning." *Journal of Genetic Psychology* 108 (1966):121-134. *G T*

In a study using sophisticated experimental apparatus and controls, three out of six Rhesus monkeys were able to recognize pleasurable responses in other monkeys: i.e., "facial expression or head movement of other monkeys receiving stimuli associated with reward." Changes in heart rates of the "perceiving" monkeys were similar to those accompanying a conditioned stimulus when they were trained.

666 Miller, Robert E.; Murphy, John V.; and Mirsky, I. Arthur. "Relevance of Facial Expression and Posture as Cues in Communication of Affect between Monkeys." *A.M.A. Archives of General Psychiatry* 1 (1959):480-488. *D G P*

Photographs of familar monkeys with calm expressions of face and body were shown to other monkeys through a one-way screen. The photographs were combined with shock in order to condition bar pressing to them. A series of tests followed for stimulus generalization, stimulus discrimination, partial extinction, and reconditioning involving photographs of familiar and unfamiliar calm or fearful monkeys. Results showed that the monkeys can discriminate between the fearful and nonfearful pictures and make more avoidance responses to the fearful pictures than to the nonfearful pictures which were originally associated with shock.

667 Miner, Horace. "Body Ritual among the Nacirema." *American Anthropologist* 58 (1956):503-507.

Describes the secret and painful body rituals of this North American culture, which has a strong preoccupation with the body and considers it ugly and disease-prone.

668 Mira, Emilio. "Myokinetic Psychodiagnosis: A New Technique of Exploring the Conative Trends of Personality." *Proceedings of the Royal Society of Medicine* 33 (1940):173-194. *D G T*

The myokinetic technique consists of having subjects draw various lines and circles in different spatial planes without being able to see. Forty lines in three planes are obtained. The test equipment and procedures, methods of evaluation, and preliminary results of its potential for diagnosing various psychiatric disorders (different organic, schizophrenic, and neurotic syndromes) are presented.

669 Mirsky, I. Arthur; Miller, Robert E.; and Murphy, John V. "The Communication of Affect in Rhesus Monkeys: I, An Experimental Method." *Journal of the American Psychoanalytic Association* 6 (1958):433-441. *G*

Monkeys were conditioned to bar-press to avoid shock when they saw a monkey through a one-way screen. This was then extinguished while they observed the other monkey. The setup was repeated, but with the stimulus monkey being shocked. The nonshocked, observing monkeys began frantically pressing the bar again, but would not do this when observing a shocked rabbit or a monkeylike object being moved agitatedly.

670 Mishra, Rammurti S. *Fundamentals of Yoga: A Handbook of Theory, Practice and Application.* New York: Julian Press, 1959. 255 pp. *D*

The author, himself a physician and practitioner of raja-yoga, discusses the potential of yoga for treatment of physical and mental distress through meditation, control of posture and breathing, relaxation of body organs, and development of body consciousness. Principles of yoga practice, body discipline, gaze training, etc., are presented within the framework of yoga philosophy.

671 Misner, William D. "The Significance of Mobility in Early Childhood: Comparison in Two American Indian Cultures." *Human Potential* 2 (1969):15-20.

Grade and high school Navaho and Hopi students were given AAHPER physical fitness tests, the California Reading Tests, and the Grace Arthur Point Performance Test and the Goodenough "Draw-A-Man" Test for nonverbal measures of I.Q. Hopi children scored consistently higher on all tests. The author posits a correlation between the Hopi's apparently higher intellectual and motor skills and the greater freedom they have to move and explore in infancy. Navaho children, in contrast, receive less attention from adults and are bound in cradleboards for the first two years of life. They apparently do not crawl or creep in infancy but go immediately from cradleboard to walking.

672 **Mittelman, Bela.** "Motility in Infants, Children, and Adults: Patterning and Psychodynamics." *Psychoanalytic Study of the Child* 9 (1954):142-177. D

A most valuable analysis of the origins, development, and motivational bases of patterns of motility, richly illustrated with observations of children in a two-and-one-half-year study and patients in long-term psychotherapy. The author discusses motility as an "urge" or "drive" in its own right; a distinct motor phase of ego development; the role of movement in other physiological and libidinal urges; how various emotions are expressed motorically at different ages; rhythmic movements in young children; individual differences in "affectomotor" behavior; and the pathological results of constricting movement. The relationship between development of prehension, posture, and locomotion and the development of self-image and the ego functions is assessed. The developmental stages and psychological implications of "aggressive motor behavior," directed outward and inward toward the self, are outlined. The article concludes with an extensive discussion of the etiology of hyper- and hypoactivity and motor pathology and the movement characteristics of various psychodiagnostic groups.

673 **Mittelmann, Bela.** "Motor Patterns and Genital Behavior: Fetishism." *Psychoanalytic Study of the Child* 10 (1955):241-263. D

The author posits that there is a general motor drive made up of "affectomotor" patterns (bodily expression of emotion), vigorous, autoerotic, rhythmic patterns; motor skills and motor phenomena indispensable to "the function of the other organs." The developmental, sexual, and psychological significance of affectomotor and rhythmic movement patterns is focused on in this paper. By means of film samples of normal and disturbed children and adults, motor patterns characteristic of specific emotions are carefully described and analyzed. The observations of adult "affectomotor" behavior and rhythmic actions are compared with early childhood patterns. An analysis of Hitler's movement as filmed during moments of extreme emotional excitement is included. Much of the paper is devoted to an analysis of how these motor phenomena may play a part in the etiology of neurotic "disturbances of the genital function," illustrating this with a description of the history, psychodynamics, and motor patterns of a few patients with fetishes.

674 **Mittelmann, Bela.** "Motility in the Therapy of Children and Adults." *Psychoanalytic Study of the Child* 12 (1957):284-319.

A recapitulation of the author's earlier writing on the "motor drive" and its importance in character formation is presented, along with a survey of psychoanalytic literature on movement behavior. Citing case examples from child therapy, the author discusses motivational aspects of play behavior, "regressive" forms of expression, ticlike movements, and diminished activity and hyperactivity, together with notes on how the therapist may respond. Motor themes (notably restriction of freedom to move and motor awkwardness) derived from reports of adult neurotic and psychotic patients are then discussed, with observations of their movement behavior.

675 **Mittelmann, Bela.** "Psychodynamics of Motility." *International Journal of Psycho-Analysis* 39 (1958):196-199.

Brief but very interesting, this article is a psychoanalytic discussion of the function of motility in normal and pathological development and its particular importance in the second year, called here the "motor phase of ego and libido development." Its importance is illustrated in descriptions of dreams and memories of patients in analysis, an early memory of motor restriction reported by Tolstoy, and experiences with children in treatment.

676 **Mobbs, N. A.** "Eye-Contact in Relation to Social Introversion/Extroversion." *British Journal of Social and Clinical Psychology* 7 (1968):305-306. *T*

In a controlled situation in which they each discussed a TAT picture, subjects classified as extroverts showed greater average duration of eye contact than those considered neutral or introverted.

677 **Montagu, Ashley.** *Touching: The Human Significance of the Skin.* New York: Columbia University Press, 1971. 338 pp. *G*

The major work on the developmental and interpersonal significance of touching. Drawing from a great deal of literature which is listed in a reference section, Montagu explores what influence "various kinds of cutaneous experiences which the organism undergoes, especially in early life, have upon its development" (p. 11). Touch in animal behavior, prenatal life, and mother-child relationships is described with care. Sex, socioeconomic, and cultural differences are assesssed. In this moving and compassionate work, patterns of touch in various cultures, particularly Balinese, Eskimo, American, Russian, and Japanese, are described. Analysis of the physiology of tactile sensation is integrated with the psychological and anthropological observations.

678 **Moore, H. T., and Gilliland, A. R.** "The Measurement of Aggressiveness." *Journal of Applied Psychology* 5 (1921):97-118. *D P T*

Two groups of men were selected according to ratings of high aggressiveness (initiative, assurance) and low aggressiveness. The high-aggressiveness group showed six eye movements away from the examiner, as compared with the low group's seventy-two. (They were to maintain eye contact throughout the mental task.) Other comparisons such as eye movements during various distractions and definiteness of response to word associations discriminated between the two groups.

679 **Moore, Joseph E.** "A Test of Eye-Hand Coordination." *Journal of Applied Psychology* 21 (1937):668-672. *D G*

The apparatus and procedure for the Moore Eye-Hand Coordination Test is described, and its reliability and correlation with another motor test are given. Designed for both children and adults, this test indicates age differences.

680 **Moreno, J. L.** "Psychodrama." In *American Handbook of Psychiatry,* vol. 2, edited by S. Arieti, pp. 1375-1396. New York: Basic Books, 1959.

A most interesting account of the development of psychodrama and the rules and techniques formulated by its founder, Moreno. As he says here, psychodrama represents "the rebellion of the suppressed actor against the word" (p. 1375). With its techniques of role playing, protagonist and auxiliary ego figures, acting out, etc., psychodrama represents one of the earliest forces toward deemphasizing the verbal and utilizing the nonverbal behavior of patients in therapy.

681 **Morris, Charles.** *Signs, Language, and Behavior.* Englewood Cliffs, N.J.: Prentice-Hall, 1946. 365 pp. [New York: George Braziller, 1955]

Research in movement behavior often becomes entangled in discussions of terms such as expressive, communicative, signal, and symbolic, and this book contains a useful analysis of the semantics involved in these debates.

682 **Morris, Desmond.** "The Feather Postures of Birds and the Problem of the Origin of Social Signals." *Behaviour* 9 (1956):75-113. *D*

An extensive discussion of the feather movements of various birds, including their function in regulating body temperature and as social signals. Social behaviors are described, notably imitation, moving close in response to a bird who rests with "fluffed" feathers, preening, and "ruffling" feathers in courtship. Primary and secondary somatic signals and "automatic signals" arising from "thwarting situations" are assessed. Included also is a short bibliography on bird ethology.

683 **Morris, Desmond,** ed. *Primate Ethology.* Chicago: Aldine Publishing Company, 1967. 374 pp. *D G P T*

A great deal of description, illustration, and analysis of body movements and facial expressions of primates are included in the following chapters: "The Facial Displays of the Catarrhine Monkeys and Apes" by J. A. R. A. M. Van Hooff, "Socio-sexual Signals and Their Intra-specific Imitation among Primates" by Wolfgang Wickler; "Allogrooming in Primates: A Review" by John Sparks; "Play Behaviour in Higher Primates: A Review" by Caroline Loizos; "Comparative Aspects of Communication in New World Primates" by M. Moynihan; "The Effect of Social Companions on Mother-Infant Relations in Rhesus Monkeys" by R. A. Hinde and Y. Spencer-Booth; "Mother-Offspring Relationships in Free-ranging Chimpanzees" by Jane van Lawick-Goodall; and "An Ethological Study of Some Aspects of Social Behaviour of Children in Nursery School" by N. G. Blurton Jones.

684 **Morris, Margaret.** *The Notation of Movement.* London: Kegan Paul, Trench, Trubner & Co., 1928. 103 pp. *D N*

A notation system developed from dance that has a six-line staff—with the upper segment for head and arms, the lower for legs and feet, and a center space for the trunk. There are symbols for direction, degree of flexion-extension, rotation, pronation-supination, breathing, turns, types of accents and transition, and facial movements.

685 **Mosher, Harris D.** "The Expression of the Face and Man's Type of Body as Indicators of His Character." *Laryngoscope* 61 (1951):1-38. *D P*

A lecture on the evolution of the anatomy of head and face from fish to man; the facial muscles of expression; and the history of physiognomy, with a note about modern research on character and body type.

686 **Mosher, Joseph A.** *The Essentials of Effective Gesture.* New York: The Macmillan Company, 1916. 188 pp. *N*

Attempting to outline the "significance of various positions and lines of movement," rather than teach specific gestures for public speaking, the author classifies types of gestures (literal versus figurative, emphatic, descriptive, locating, expressive or actions, and expressive of mental or emotional states). He describes three phases of the gesture, qualities desired, and aspects of frequency, facial expression, and position. Delineating planes and hand positions with a letter notation, he illustrates how certain spoken ideas and feelings may be accompanied by specific sequences of movement.

687 **Mühl, Anita M.** "Automatic Writing as an Indicator of the Fundamental

Factors Underlying the Personality." *Journal of Abnormal and Social Psychology* 17 (1922-1923):162-183. *D*

An incredible report about two different women who showed "automatic writing": i.e., they would draw or write often without awareness and in script not their own, as people other than themselves. One woman composed elaborate stories; the other's writing was a drama about different personalities, each with different writing, who interacted with her and with each other. There are descriptions of this woman's expressions and movements as the different characters "spoke" through her writing.

688   **Munn, Norman L.** "The Effect of Knowledge of the Situation upon Judgment of Emotion from Facial Expressions." *Journal of Abnormal and Social Psychology* 35 (1940):324-338. *P T*

Ninety subjects were shown candid photographs of individuals in real and dramatized situations. First they were shown only the face, and a week later the entire picture; both times they wrote down what emotion they thought was being experienced. A second group of sixty-five saw the entire picture for all and were given a list of specific emotion terms to check off. Both groups showed high agreement. In some cases, seeing the entire picture did not increase agreement achieved by seeing just the face; in other cases it did.

689   **Nadel, Myron H., and Nadel, Constance G.** *The Dance Experience: Readings in Dance Appreciation.* New York: Praeger Publishers, 1970. 388 pp. *N*

A collection of articles by dancers, choreographers, dance critics, and educators that surveys the field of dance—its nature, forms and relation to other arts, the choreographic process, types of dance notation, dance aesthetics, criticism and education, and special problems in dance. Of particular note here are the following articles: "A Prolegomenon to an Aesthetics of Dance" by Selma Jeanne Cohen; "Virtual Powers" by Susanne K. Langer; "Symbolic Forms of Movement: Dance" by Eleanor Methany with Lois Ellfeldt; "Phenomenology: An Approach to Dance" by Maxine Sheets; "Education through Dance" by Margaret H'Doubler; and "Psychomotor Function as Correlated with Body Mechanics and Posture" by Lulu E. Sweigard.

690   **Nakajima, Bun.** *Japanese Etiquette.* Tokyo: Japan Travel Bureau, 1957. 222 pp. *D P*

This book for foreign visitors presents and illustrates a great deal of information on how to sit, stand, and move at Japanese social gatherings and ceremonies.

691   **Narodny, Ivan.** *The Dance.* New York: National Society of Music, 1916. 284 pp. *D P*

An extensive though dated survey of historical, ethnic, and theatrical dance, its aesthetics, origins, and interpretation. Of particular note are the descriptions of American Indian ceremonies and folk dances of Central and Western Europe.

692   **Needles, William.** "Gesticulation and Speech." *International Journal of Psycho-Analysis* 40 (1959):291-294.

The "regressive element" of gesticulations, which may replace speech when one is under emotional stress, is analyzed.

693   **Nissen, Henry W.** "A Field Study of the Chimpanzee: Observations of Chimpanzee Behavior and Environment in Western French Guinea." *Comparative Psychology Monographs,* Serial no. 36, 8 (1931):1-22. *P T*

This early field study of about twenty-five groups of chimpanzees includes observation records and discussion of their locomotion and activity patterns and social behaviors such as gesticulations and vocalizations.

694   **Nony, Camille.** "The Biological and Social Significance of the Expression of the Emotions." *British Journal of Psychology* 13 (1922):76-91.

A theoretical paper on the nature and function of emotional reactions (physiological, motoric, etc.), with a critique of prior theories and an analysis of the necessary conditions for emotional expression to be a language.

695   **North, Marion.** *Personality Assessment through Movement.* London: Macdonald & Evans, forthcoming.

According to a prepublication announcement, the author describes how to observe, notate, and analyze an individual's movement patterns and also presents movement assessments and case studies of children and adults. She has developed Laban's approach to movement analysis for some years and applied it to personality assessment.

696   **Norum, Gary A.; Russo, Nancy J.; and Sommer, Robert.** "Seating Patterns and Group Task." *Psychology in the Schools* 4 (1967):276-280. *T*

Two studies, one with schoolchildren and one with college students, showed that when the situation is cooperative subjects elect to sit next to or closer to each other than when it is competitive, and they sit farthest apart when working individually.

697   **Oesterley, W. O. E.** *The Sacred Dance: A Study in Comparative Folklore.* (Cambridge University Press, 1923; reprint ed., Brooklyn, N.Y.: Dance Horizons, 1968). 234 pp.

A minister's account of the origins and purposes of religious rites and dances, particularly in ancient cultures. This book is very dated and biased with regard to ethnological dance of existing cultures, but its descriptions and interpretations of rituals and dances culled from the Old Testament and from Greek and Roman literature are notable.

698   **Olson, Willard C.** "The Incidence of Nervous Habits in Children." *Journal of Abnormal Psychology* 25 (1930):75-92. *T*

The "nervous habits" of nursery and elementary schoolchildren were systematically observed. Behaviors such as "clenching the fist," "rubbing eyes," etc., were grouped into nine categories, and the incidence of "oral habits" was evaluated in detail, indicating no age differences, significantly higher incidence in girls, and individual differences. An analysis of family characteristics, comparisons between siblings and between schoolchildren in proximity to each other, and correlations with physical characteristics suggest that the incidence of oral habits is related to neuroses in the family, length of breast feeding, imitation of others, and nutritional factors.

699   **Osgood, Charles E.** "Dimensionality of the Semantic Space for the Communication Via Facial Expressions." *Scandanavian Journal of Psychology* 7 (1966): 1-30. *D T*

This factor analysis of students' judgments of the facial expressions of student actors yielded three dimensions: pleasantness, activation, and control, as well as several clusters or types of expression. A schema for developmental stages of emotional expression based on the model derived and a discussion of the culture-specific nature of emotional expression are also presented.

700   **Overby, Charles M.** "Experimental Cybernetic Analysis of Hand Action."

Ph.D. dissertation, University of Wisconsin, 1965. 205 pp. (Datrix order no 65-14920)

Research equipment and a computer program were developed to study prehension—in this case, accuracy on a tracking task varying handedness, finger-thumb apposition, visual and proprioceptive feedback, and task difficulty. Varying of the tests, instructions to the subjects, scoring, and storing of the results were accomplished by computer. The author presents some of the results: e.g., in a certain task left-hand performance did not differ from right-hand.

701   **Oxendine, Joseph B.** *Psychology of Motor Learning.* New York: Appleton-Century-Crofts, 1968. 366 pp. *D G T*

Motor learning is a subject not covered in the present bibliography, and the reader is referred to books such as this one, which has an extensive bibliography and review of the literature on motor learning. It summarizes traditional theories of learning and motivation in relation to motor learning and performance and has sections on developmental and sex differences, the relation of motor skill to intelligence, kinesthesis and motor performance, laterality, and research on movement speed.

702   **Oxenford, Lyn.** *Design for Movement: A Textbook on Stage Movement.* New York: Theatre Arts Books, 1951. 96 pp. *D*

Somewhat sketchy but interesting discussions and exercises on movement for actor training. The individual's "habitual physical attitude" and characteristic movement patterns in terms of patterns of space, force, time, and amount of space used, and what these factors convey about character and emotional state are discussed. There are observations on cultural differences in body movement, stylized movement, and analysis of group movement, as well as a chapter on the physical movements of religious ritual.

703   **Paget, Sir Richard.** *Human Speech: Some Observations, Experiments, and Conclusions as to the Nature, Origin, Purpose and Possible Improvement of Human Speech.* New York: Harcourt, Brace and Company, 1930. 360 pp. *D G P T* [New York: Humanities Press, 1963]

Cited here for its discussion of how speech originated from pantomimic gestures of mouth and total body and how the mouth movements often have a "pantomimic relation" to the idea expressed.

704   **Pease, Esther,** ed. *Compilation of Dance Research: 1901-1964.* Washington, D.C.: American Association for Health, Physical Education, and Recreation, National Section on Dance, 1964. 53 pp.

A 704-title bibliography of graduate level research in dance (performance, education, ethnic, historical, aesthetics), dance therapy, and other fields (psychology, religion, physical education, motor skills), with a subject index.

705   **Pennington, Jo.** *The Importance of Being Rhythmic.* New York and London: G. P. Putnam's Sons, 1925. 142 pp. *D P*

Of historical interest, this book is based on the principles of "eurhythmics" of Jacques-Dalcroze. Dalcroze considered training in rhythm (through music and movement) as a crucial part of education and mental, physical, and emotional development. For example, attention and inhibition could be developed through movement rhythm exercises. This book surveys the development of eurhythmics and its applications to drama and dance.

706   **Perls, Frederick; Hefferline, Ralph F.; and Goodman, Paul.** *Gestalt Therapy:*

Excitement and Growth in the Human Personality. New York: Dell Publishing Co., 1951. 470 pp *D* [New York: Julian Press, 1969]

This book is both treatise and treatment modality in that it presents an elaborate theory of normal and disturbed behavior derived from Gestalt psychology, while having the reader perform specific exercises as he reads to help him understand and personally change. Specific sections relevant to the subject of movement behavior are on body awareness (pp. 82-94); anxiety as the experience of respiratory and chest constriction (pp. 128-135); the significance of self-directed movements and muscular tensing to suppression of impulses, psychological interpretation of specific areas of tension, and the reversal of these processes and expression of emotion (pp. 161-188, 204-205, and 311-312). The reactions of individuals who performed the exercises are presented, providing introspective accounts of experiences of specific muscular tensions and movements.

•707 **Pesso, Albert.** *Movement in Psychotherapy: Psychomotor Techniques and Training*. New York: New York University Press, 1969. 221 pp.

The author describes techniques of "psychomotor therapy" he and his wife have developed in work with groups in mental hospitals and treatment centers. A progression is described from experience of the "species stance" and "reflexive, voluntary, and emotional motor impulses" to a reenactment of specific emotional situations and early relationships with group members representing negative and positive aspects of significant early figures. A clear and comprehensive presentation of the therapeutic use of movement and movement correlates of personality and emotion.

708 **Phillips, John C.** "A Study of Motor Functions of Selected Groups of Schizophrenics." Ph.D. dissertation, Temple University, 1954. 100 pp. (Datrix order no. 60-2621)

A total of 160 men between twenty and forty years old—40 acutely ill schizophrenics, 40 schizophrenics in remission, 40 chronically ill schizophrenics, and 40 normal subjects—were given a series of eight motor tests such as the MacQuarrie Tapping and Dotting Test, rail walking, strength of grip, etc. Analysis of the results showed that the acute and chronic patients were much alike, and that the schizophrenics in remission compared to the normal subjects in motor performance on these tests.

709 **Pickersgill, M. Gertrude.** *Practical Miming*. London: Sir Isaac Pitman & Sons, 1936. 117 pp. *D P*

Though dated and simplistic, there are some interesting details in this collection of exercises for the mime student: how to express certain emotions, group relationships, and "stylized gestures" evolved from sixteenth- and seventeenth-century Italian mime.

710 **Plutchik, Robert.** "The Role of Muscular Tension in Maladjustment." *Journal of General Psychology* 50 (1954):45-62.

A cogent review of literature on muscle tension which supports the author's hypothesis that body expressions and postures are aspects of personality, and that chronic muscle tension is related to intrapsychic conflict and frustration and to the neuroses and psychoses.

711 **Plutchik, Robert.** *The Emotions: Facts, Theories, and a New Model*. New York: Random House, 1962. 204 pp. *D T*

Following a review of problems in studying emotions and major theories of emotions, the author proposes specific postulates about the nature of emotions. He argues for eight primary dimensions and their corresponding "prototypic patterns" of behavior and. includes description of bodily changes and motor expression of the emotions. Among studies reported that support his theories is one involving judgment of facial expression.

712   Polak, Paul R.; Emde, Robert N.; and Spitz, René A. "The Smiling Response to the Human Face: I, Methodology, Quantification, Natural History." *Journal of Nervous and Mental Disease* 139 (1964):103-109. *G T*

A mode of eliciting and rating the smiling response without artificial appartus is described. Developmental stages in the latency, duration, and intensity of this "nonautomatic" smiling are assessed in children ages two to eight months.

713   Pollack, Max; and Krieger, Howard P. "Oculomotor and Postural Patterns in Schizophrenic Children." *A.M.A. Archives of Neurology and Psychiatry* 79 (1958):720-726. *D G T*

The eye movements of fifteen schizophrenic children (ages seven to nine), seven children with behavior disorders, and nine normal children were examined with an "optokinetic drum" apparatus. A number of the schizophrenic children showed involuntary head turning during visual stimulation, "inability to dissociate head movement from eye movement," as well as deviant postural reactions and difficulty in crossing the body midline and performing complex movements.

714   Pollenz, Phillippa. "Methods for the Comparative Study of the Dance." *American Anthropologist* 51 (1949):428-435. *D N*

A discussion of methods of recording ethnological dance (with word descriptions, track drawings, stick figures, and Labanotation) is presented, with references to specific anthropological studies using dance recording.

715   Ponder, Eric, and Kennedy, W. P. "On the Act of Blinking." *Quarterly Journal of Experimental Physiology* 18 (1927):89-110. *G T*

A very interesting study of the emotional and physical factors in blinking. The authors recorded blinking patterns of individuals in emotional excitement, witnesses under cross-examination, people in reading rooms, passengers on streetcars, and subjects who drank a lot of alcohol. Individual consistency and sex differences in "interblink periods" and the relation of blinking to "mental tension" are discussed. Also, normal blink rates were found in congenitally blind people. A number of experimental studies of physical factors in blinking are reported.

716   Prechtl, H. F. R. "The Directed Head Turning Response and Allied Movements of the Human Baby." *Behaviour* 13 (1958):212-242. *D T*

The side-to-side head movements of young infants were studied while they were feeding, lying in a crib, or being stimulated in an experimental situation. Neurological tests were given along with qualitative and quantitative analyses of the head movement, especially the "rooting" response to stimulation of mouth or cheek. The development of head orienting and variations in response threshold and intensity are discussed in relation to physiological processes.

717   Prechtl, H. F. R., and Stemmer, Ch. J. "The Choreiform Syndrome in Children." *Developmental Medicine and Child Neurology* 4 (1962):119-127. *D G*

Fifty hyperactive children with poor school performances were given extensive examinations (EMG, EEG, blood tests, case histories, etc.). They all showed slight, arhythmic, jerky movements that the authors called part of the "choreiform syndrome." Test results are presented, including evidence of minimal brain damage.

718 **Preston, Valeria.** *A Handbook for Modern Educational Dance.* London: Macdonald & Evans, 1963. 187 pp. *D N T*

The concepts and movement notation of Rudolf Laban, presented as they are here in the form of sixteen "basic movement themes" (e.g., body awareness, awareness of weight and time, orientation in space, group formation, etc.), can be seen as analgous to music theory and notation. Primarily a textbook and curriculum for the dance teacher, this work says a lot about how Laban and his colleagues observe, interpret, and record movement and develop dances and movement training through principles of movement rather than imitation of the teacher.

719 **Preston-Dunlap, Valerie.** "A Notation System for Recording Observable Motion." *International Journal of Man-Machine Studies* 1 (1969):361-386. *D N*

A presentation of OMDR (Observable Motion Data Recording), which combines Laban's effort analysis and Kinetography, or Labanotation. The author discusses the need for a notation and language of movement that does justice to the refinements of a skilled work, the spirit of a dance, and the logic required in movement research. A description with illustrations of each of the four possible columns of OMDR is presented: Kinetography Laban, or detailed notation for reconstruction; effort symbols for space, time, pressure, and "flow" variations; "motif writing," or the essential features of a movement recorded in Kinetography symbols; and linear effort graphs, or effort variations on a time scale. There is a discussion of the uses of OMDR, with stress on job skill analysis.

720 **Preston-Dunlop, Valerie.** *Practical Kinetography Laban.* London: Macdonald & Evans, 1969. 216 pp. *N* [Brooklyn, N.Y.: Dance Horizons, 1970]

A concise, readable presentation of Kinetography Laban movement notation (Labanotation), which has sections on ways to record patterns of movement intensity and dynamics.

721 **Rafi, Amin Abi.** "Motor Performance of Certain Categories of Mental Patients." *Perceptual and Motor Skills* 10 (1960):39-42. *T*

Performance on a series of simple motor tasks was shown to discriminate between chronic schizophrenics, "mild schizophrenics," and normal controls.

722 **Rand, George, and Wapner, Seymour.** "Postural Status as a Factor in Memory." *Journal of Verbal Learning and Verbal Behavior* 6 (1967):268-271.

Subjects who learned nonsense syllables under congruent conditions (e.g., learn standing up, relearn standing or learn lying supine, relearn supine) recalled somewhat better than those learning in incongruent situations (e.g., learning erect, relearning supine).

723 **Redfern, Betty.** *Introducing Laban Art of Movement.* London: Macdonald & Evans, 1965. 32 pp.

A short theoretical work on the application of Rudolf Laban's "effort analysis" of movement to principles of dance education, assessment of job skill in industry, use of movement in psychodiagnosis and therapy, and movement principles in the theater and in recreational dance. While familiarity with effort analysis helps, the book is a good exposition of movement concepts emerging from Laban's work and that of his colleagues at the Art of Movement Centre in England.

724   **Reece, Michael M., and Whitman, Robert N.** "Expressive Movement, Warmth and Verbal Reinforcement." *Journal of Abnormal and Social Psychology* 64 (1962):234-236. *T*

Subjects were asked to free-associate for fifteen minutes. Those with a "warm" experimenter (who leaned forward, smiled, and looked at the subject) verbalized more than those with a "cold" experimenter (who leaned away, looked around, and drummed his fingers). Verbal reinforcement alone increased the number of plural nouns as intended, but not the total number of words.

725   **Reich, Wilhelm.** *The Function of the Orgasm.* Translated by T. P. Wolfe. New York: The Noonday Press, 1942. 368 pp. *D G T* [New York: Farrar, Straus & Giroux (The Noonday Press), 1961]

Widespread rejection of Reich's "Orgone" theory and his later experiments and writing seems to have brought with it an obscuring or dismissal of his writing on the importance of muscular attitudes, breathing, and patterns of tension in character analysis and personality, but very recently there is evidence of a renewed interest in Reich's work. Sections of this book are eminently valuable and historically important to the study of movement behavior. These sections are the description of mature sexual intercourse (pp. 72-87), notes on chronic muscular tension and anxiety, anger and excitation (pp. 240-242), theories concerning the origin and significance of specific types of muscular tension, an elaborate case description of direct analysis of muscular attitudes ("muscular vegeto-therapy"; pp. 270-292), and the significance of pathological breathing patterns and "dead," immobile areas of the body (pp. 292-320).

726   **Reich, Wilhelm.** *Character-Analysis.* 3rd ed. Translated by T. P. Wolfe. New York: Farrar, Straus & Giroux (The Noonday Press), 1949. 516 pp. *D*

In this classic psychoanalytic work on characterology and "resistance analysis," Reich makes the transition to defining, manifestations of defense mechanisms, "character armor," psychosexual fixation and personality characteristics in the body musculature, posture and manner of walking, gesturing, and breathing. The clinical descriptions and interpretations of patterns of movement and "chronically fixed muscular attitudes" are brilliant and provocative, particularly the references to the patient with an "aristocratic character" and the schizophrenic patient, the observations on muscular rigidity or flaccidity in various character types and schizophrenics, and the descriptions of how patients inhibit excitation and sexuality with muscular rigidity and avoid emotional contact with "ungenuine behavior." Reich discusses the function of the character armor and its treatment. He formulates a theory of character formation and sexuality, for which patterns of movement and muscular tension are an integral part.

727   **Reiter, Paul J.** "Extrapyramidal Motor-Disturbances in Dementia Praecox." *Acta Psychiatrica et Neurologica* 1 (1926):287-309. *P*

Ten medical and psychiatric case histories are presented involving patients with dementia praecox who showed marked motor disturbances. The types of abnormal movements observed and some post-mortem evidence of gastrointestinal disorders and brain lesions are described.

728   **Renneker, Richard.** "Kinesic Research and Therapeutic Processes: Further Discussion." In *Expression of the Emotions in Man*, edited by P. H. Knapp, pp. 147-160. New York: International Universities Press, 1963.

Drawing on four years of research on movement in psychotherapy, the author discusses guidelines for observing, equipment, research goals, and hypotheses about the function and nature of body movement.

729 Reymert, Martin L., and Speer, George S. "Does the Luria Technique Measure Emotion or Merely Bodily Tension? A Re-evaluation of the Method." *Character and Personality* 7 (1938-39):192-200. *T*

Following a review of research using Luria's techniques for measuring motor tension, the authors report a study in which the records of motor tensions of children responding to "emotionless" questions while using the Luria apparatus were the same as those Luria obtained from subjects under emotional stress.

730 Rhodes, Adele. "A Comparative Study of Motor Abilities of Negroes and Whites." *Child Development* 8 (1937):369-371. *T*

Eighty Negro children from two to five years old were given a series of structured motor tests, and their scores were compared with those of white children from another study. Both groups were very similar in level of ability. Two motor factors found, "general motor maturity" and "carefulness," paralleled those founded in a previous study by Goodenough and Smart.

731 Richards, T. W., and Newbery, Helen. "Studies in Fetal Behavior: III, Can Performance on Test Items at Six Months Postnatally Be Predicted on the Basis of Fetal Activity?" *Child Development* 9 (1939):79-86. *T*

An initial study of twelve infants indicating a positive relationship between performance on the Gesell schedule at six months and the mothers' reports of fetal activity.

732 Richards, T. W., and Simons, Marjorie Powell. "The Fels Child Behavior Scales." *Genetic Psychology Monographs* 24 (1941):259-309. *T*

Behavior scales developed for nursery school children are described. Among thirty personality traits rated are scales for "Frequency of Gross Activity," "Physical Apprehensiveness," and "Vigor of Activity." Definitions of each item, instructions for scoring, high observer agreement, certain marked intercorrelations, and data on validity of the scales are discussed.

733 Riemer, Morris D. "The Averted Gaze." *Psychiatric Quarterly* 23 (1949):108-115.

Case summaries of six patients in psychotherapy who showed "fixed averted gaze" that are presented indicate that these people were severely estranged from their parents; their parents were rarely warm or affectionate and rarely looked at them; the parent of the same sex was weaker or more remote; the patients themselves are very "moody" and resentful, detached and afraid of rejection. The author considers a fixed averted gaze pathognomonic of schizophrenia.

734 Riemer, Morris D. "Abnormalities of the Gaze: A Classification." *Psychiatric Quarterly* 29 (1955):659-672.

The origins and defensive functions of abnormal gaze patterns and accompanying face and body mannerisms are discussed, and six types are described and interpreted in the order of their severity: excessive blinking, the depressed look, the dramatic gaze, the guarded gaze, the absent gaze, and the averted gaze.

735 Robins, Ferris and Jennet. *Educational Rhythmics for Mentally and Physically Handicapped Children.* New York: Association Press, 1968. 239 pp. *D P*

Essentially a book of structured movement exercises and activities for the handicapped child (retarded, blind, deaf, and those with motor disability), which educates them in concepts and capacities such as coordination, concentration, time, and writing. There are brief references to their body movement characteristics and to ways of assessing improvement. It is richly illustrated with photographs and drawings.

736   **Robson, Kenneth S.** "The Role of Eye-to-Eye Contact in Maternal-Infant Attachment." *Journal of Child Psychology and Psychiatry* 8 (1967):13-25.

A beautifully written review of the literature on attention, visual focus, and the importance of eye-to-eye contact in the mother-child relationship in the first six months. Types of eye contact and gaze aversion are discussed, along with clinical notes on abnormal eye-contact behavior and psychopathology.

737   **Rolf, Ida P.** "Structural Integration: Gravity, An Unexplored Factor in a More Human Use of Human Beings." *Systematics* 1 (1963):66-83. *P* (Copies from the Guild for Structural Integration, 1776 Union Street, San Francisco, Calif. 94123)

The author discusses the theoretical bases and effectiveness of her "structural integration" methods of improving body structure and posture. The work is put within a psychological framework as she discusses how types of posture and chronic displacement of body parts represent "scars" of emotional and physical trauma.

738   **Rorschach, Hermann.** *Psychodiagnostics.* (Verlag Hans Huber, 1942; reprint ed., New York: Grune & Stratton, 1969). 228 pp. *T*

The relation between actual physical movement characteristics and the "M" response, or perceived movement on the Rorschach test of personality, is a controversial one in current research; but Rorschach himself does not hesitate to make hypotheses and interpretations of it. He assesses the movement characteristics of those with a predominance of M responses versus those with C responses (p. 98); the relation between M responses and the degree of physical activity (pp. 78-83); M responses of manic individuals versus schizophrenics in motor excitement (p. 161); and the kinds of movement perceptions in the Rorschach and what they imply about the personality (e.g., those who see extension movements tend to be more active and assertive).

739   **Rosen, Elizabeth.** *Dance in Psychotherapy.* New York: Teachers College Press, Columbia University, 1957. 178 pp. *P*

A presentation of the theory and practice of dance therapy with schizophrenic and manic-depressive psychiatric patients. Descriptions of individual patients at Hillside Hospital and Manhattan State Hospital, their participation in dance therapy sessions, and their reactions to them are presented.

740   **Rosenberg, B. G., and Langer, Jonas.** "A Study of Postural-Gestural Communication." *Journal of Personality and Social Psychology* 2 (1965):593-597. *D T*

Subjects were asked to judge drawings of stick figures in various postures according to several dimensions (feeling, stability, vertical direction, horizontal direction, chromatic color, and achromatic color). Judgments were highly consistent.

741   **Rosenfeld, Howard M.** "Instrumental Affiliative Functions of Facial and Gestural Expressions." *Journal of Personality and Social Psychology* 4 (1966): 65-72. *T*

Subjects asked to approach a stranger and seek her approval and liking showed more smiling, head nods, gesticulations, and less self-manipulation than those asked to avoid approval. In a second study, subjects assessed to have a higher need for approval smiled more.

742   **Rosenfeld, Howard M.** "Approval-Seeking and Approval-Inducing Func-

tions of Verbal and Nonverbal Responses in the Dyad." *Journal of Personality and Social Psychology* 4 (1966):597-605. *T*

One member of each experimental dyad was instructed to act toward the other as he would if he liked the person (called the "approval-seeking condition," or AS) or as he would if he did not find the other attractive ("approval-avoiding," or AA). AS subjects were twice as active, smiled, gesticulated and head-nodded more, and spoke longer. Some sex differences in AS behavior are noted. The attitudes of the "naive" subjects toward the ones acting are analyzed.

743 **Rosenfeld, Howard M**. "Nonverbal Reciprocation of Approval: An Experimental Analysis." *Journal of Experimental Social Psychology* 3 (1967):102-111. *G T*

Interviewers asked questions of students and showed approval (smiling, head nods, verbal acknowledgment, gesticulation), disapproval (frowns, head shakes indicating "no", verbal disparagement), or no response to their answers. Students smiled, head-nodded more, and showed less self-manipulatory activity in relation to the approving interviewers.

744 **Rowen, Betty.** *Learning Through Movement.* New York: Teachers College Press, Columbia University, 1963. 77 pp. *D*

How the author uses dance and movement with schoolchildren to develop group interaction, creativity, and appreciation of literature and to enhance classroom subjects such as vocabulary drill, social studies, arithmetic, and science.

745 **Ruckmick, Christian A.** "A Preliminary Study of the Emotions." *Psychological Monographs* 30 (1921):30-35. *P T*

Observers judged thirty-four photographs of facial expressions, and the lists of their interpretations are printed here. Tentative observations on the degree of agreement; effects of mood and temperament differences on judgment; and variations in ways of interpreting are noted.

746 **Ruckmick, Christian A.** *The Psychology of Feeling and Emotion.* New York: McGraw-Hill Book Company, 1936. 529 pp. *D P T*

This contains an excellent brief review of the history of studies of facial expression, from Aristotle and literature on the interpretation of physiognomy to studies of the physiology of expression to 1930's studies of observer judgment of facial expression. There is also an interesting section on early instruments for measuring bodily changes accompanying emotion, among them breathing patterns and muscle movements. Both sections are illustrated with photographs and drawings.

747 **Ruesch, Jurgen.** *Disturbed Communication: The Clinical Assessment of Normal and Pathological Communicative Behavior.* New York: W. W. Norton & Company, 1957. 337 pp.

In this book are some brief descriptions of abnormal movement patterns of schizophrenic patients and a long inventory for observation of communicative behavior, which has sections on movement disturbances.

748 **Ruesch, Jurgen, and Kees, Weldon.** *Nonverbal Communication: Notes on the Visual Perception of Human Relations.* Berkeley, Calif.: University of California Press, 1956. 205 pp. *D P*

Following a review of communication theory and biological and cultural determinants of nonverbal communication, the authors present a photographic essay to illustrate how people communicate through action, gesture, body position, and interpersonal spacing, as well as by signs, environmental effects, objects, and props. There are also a chapter on disturbances in nonverbal communication and psychopathology and concluding remarks "toward a theory of nonverbal communication."

749   Sachs, Curt. *World History of the Dance.* Translated by B. Schönberg. New York: Crown Publishers (by arrangement with W. W. Norton & Company), 1937. 469 pp. *D P* [New York: W. W. Norton & Company, 1963]

The classic work on cross-cultural analysis of the dance and the history of European dance, this book is unique among books on dance because of its approach and organization. Rather than discussing one culture at a time, it analyzes dances by their movement characteristics and themes (e.g., convulsive dances, expanded or closed dances) and the situations and themes they are addressed to (e.g., war, initiation, fertility, funeral, medicine, and animal dances).

750   Sainsbury, Peter. "The Measurement and Description of Spontaneous Movements Before and After Leucotomy." *Journal of Mental Science* 100 (1954):732-741. *G T*

Films and EMG recordings were made of forearm activity in ten psychiatric patients before and after they received prefrontal leucotomies or topectomies. The patients were filmed while alone and during an interview to study patterns of "autistic" (or self-related) versus communicative gestures. The EMG recordings correlated with the observations from film, and the number of movements tended to be individually characteristic. In comparison with normal subjects, the patients had more self-directed movements. Also, their autistic and communicative gestures and muscle tension decreased after the operation. Notably an individual's gestures did not disappear but decreased in number and intensity.

751   Sainsbury, Peter. "A Method of Measuring Spontaneous Movements by Time-Sampling Motion Pictures." *Journal of Mental Science* 100 (1954):742-748. *T*

Rates of body movements derived from time-sampling with film correlated well with the total number of movements over a longer period recorded electromyographically. Also, the rate of tics determined from short film samples correlated with that derived from continuous direct observation. The film apparatus is described.

752   Sainsbury, Peter. "Gestural Movement during Psychiatric Interview." *Psychosomatic Medicine* 17 (1955):458-469. *G T*

Heart-rate measures and EMG recordings of muscle potentials in the arms were taken of twelve patients in psychiatric interviews and were synchronized with speech. A significant correlation was found between an increase in heart rate and gestures and affectively stressful topics.

753   Sainsbury, Peter, and Gibson, J. G. "Symptoms of Anxiety and Tension and the Accompanying Physiological Changes in the Muscular System." *Journal of Neurology, Neurosurgery and Psychiatry* 17 (1954):216-224. *G T*

EMG recordings of muscle activity during an interview were made of thirty neurotic patients with marked anxiety. Upon comparing the patients' introspective reports of pain and tension with the EMG recordings, high correspondence was found between reports of pain in a specific area (e.g., the head) and a high degree of muscle tension in that area, and high tension in a number of areas accompanied generalized symptoms.

754 **Saitz, R. L., and Cervenka, Edward J.** "Handbook of Gesture: Colombia and the United States." In *Approaches to Semiotics*. The Hague: Mouton Publishers, forthcoming.

This is a revised and enlarged edition of a work formerly entitled "Colombian and North American Gestures: An Experimental Study" (Bogota, Colombia: Centro Colombo-Americano, 1962), which is now unavailable.

755 **Saunders, Eleanora B., and Isaacs, Schachne.** "Tests of Reaction-Time and Motor Inhibition in the Psychoses." *American Journal of Psychiatry* 86 (1929): 79-112. *T*

In the reaction-time tests, psychotic patients differed only in a tendency for anticipatory reactions; in the test for motor control (writing slowly), psychotic patients had more difficulty moving slowly than normal subjects.

756 **Schachtel, Ernest G.** "Projection and Its Relation to Character Attitudes and Creativity in the Kinesthetic Responses Contributions to an Understanding of Rorschach's Test: IV." *Psychiatry* 13 (1950):69-100.

In this excellent article Schachtel develops and elucidates Rorschach's interpretations of the movement (M) response to the inkblots. The M response is considered here a projection of the subject's kinesthetic perception and inhibited, habitual neuromuscular "sets" and postures. Schachtel argues that those M responses which are personally significant and from which one may interpret basic attitudes and self-image are those which recur in some way and are unusual.

757 **Schaffer, H. R., and Emerson, Peggy E.** "Patterns of Response to Physical Contact in Early Human Development." *Journal of Child Psychology and Psychiatry* 5 (1964):1-13. *T*

A comparison of two groups of infants, those reported to reciprocate and enjoy cuddling and being held and those who actively resisted it. Little difference was found in the formation of interpersonal attachments or in the mothers' preferred modes of interaction; however, "noncuddlers" were more active and had more rapid motor development.

758 **Schaffner, Bertram,** ed. *Group Processes: Transactions of the First Conference* New York: Josiah Macy, Jr., Foundation, 1954. 334 pp. *D P*

An accomplished group of researchers from a number of disciplines contributed to this conference. The format included brief presentations followed by group discussion that often diverged from the initial topic. This volume is dominated by ethological analysis of animal behavior. Of note here are the descriptions of mother-infant interaction at birth in animals and humans initiated by Dr. Blauvelt's paper; the analysis of "instinctive movement patterns" in animals by Konrad Lorenz; and the discussion of animal movements and postures initiated by N. Tinbergen.

759 **Schaller, George B.** *The Mountain Gorilla: Ecology and Behavior.* Chicago: University of Chicago Press, 1963. 431 pp. *D G P*

An exhaustive study of the environment, physical attributes, individual and social behaviors, etc., of the mountain gorilla, well illustrated and carefully described. Of note here are the descriptions of gorilla postures, locomotion patterns, grooming behavior, facial expressions in various contexts, and displays such as the chest-beating behavior. Interaction patterns and dominance, play, mating activity, etc., are described in detail, including behaviors and actions of various pairs: male-female, mother-infant, infant-infant, etc.

760 **Scheflen, Albert E.** "Communication and Regulation in Psychotherapy."
Psychiatry 26 (1963):126-136.

A brief rationale for researching psychotherapy with the context-analysis method is given. Then a specific type of behavior called "regulatory communication"
is described and illustrated with four examples from psychotherapy films. The
examples show that regulatory communication is largely kinesic: e.g., one therapist grasped the patient's hand to "monitor her tendency to become incoherent."
In another case, mother and son averted gaze when discussing being home alone
and held gaze while discussing being out together. Close analysis showed the
mutuality and complimentarity of the "monitoring" and how it occurs in oscillat-
ing sequences that regulate the "distance" between members. A discussion of the
cultural basis and the development of regulatory communication and its function in controlling deviation, pace, or relationships in interaction is presented.

761 **Scheflen, Albert E.** "The Significance of Posture in Communication Systems." *Psychiatry* 27 (1964):316-331. *D*

Estimating that there are "no more than thirty traditional American gestures"
and many fewer postural configurations. Scheflen extensively describes and illustrates the latter as studied with context analysis within a cultural framework. In
discussing how movement demarcates phases of interaction, specific eye, head,
and hand movements are described as indicators of the ends of statements; maintained head positions as defining the making of a point; gross postural shifts
as delineating the point of view taken; and changes in location as demarcating
the totality of one's position. With remarks about how these vary by sex, ethnic
group, role, or context, three specific postural relationships are described: (1)
inclusiveness-noninclusiveness of the positions and delimiting the group, (2)
vis-à-vis body orientation and reciprocal relationships, and (3) congruence-non-
congruence of body extremities and similarity of views, roles, or status.

762· **Scheflen, Albert E.** "Quasi-Courtship Behavior in Psychotherapy." *Psychiatry* 28 (1965):245-257. *D*

An article rich in observations of body movement, systematically analyzed
through context analysis and clearly placed within a theoretical framework.
Scheflen defines three stages of middle-class American courtship behavior, particularly the movement aspects of (1) readiness, (2) positioning, and (3) actions
of appeal. He discusses behavioral qualifiers that say "Don't take the sexual
behavior literally" and "decourting" actions that regulate it. The occurrence and
function of quasi-courting in psychotherapy, with specific illustrated examples,
and deviations in this behavior as it relates to psychopathology are discussed.
A theory of the function of quasi-courting in "maintaining a favorable range of
relatedness" in group interaction is presented.

763 **Scheflen, Albert E.** "Natural History Method in Psychotherapy: Communi-
cational Research," In *Methods of Research in Psychotherapy*, edited by L. A.
Gottschalk and A. H. Auerbach, pp. 263-289. New York: Appleton-Century-
Crofts, 1966. *D*

One of a number of articles in which Scheflen outlines and defends a systematic,
naturalistic approach to the study of communication and group interaction. To
show how context analysis is done, he uses an example from a psychotherapy
film that consists largely of kinesic or movement behavior. The result is a very
clear presentation of how behavioral units are abstracted, how they are related,
how the communication pattern or program is delineated, and how the "meaning" of the movement behavior is determined within this approach.

764 **Scheflen, Albert E.** *Communicative Behavior and Meaning.* New York: Doubleday & Company, forthcoming. *D*

Theoretical papers on communication, semiotics, and the analysis of meaning. Some of these include parts of the author's previous publications, and all are rich in references to kinesic behavior. Of particular note here are sections on the "representational use of nonlanguage behavior," the "bowl" gesture and its meaning, definitions of behavioral units, patterns of movements and postures in psychotherapy, behavioral programs, and kinesic styles.

765 **Scheflen, Albert E.** *Kinesics and Social Order* [tentative title]. Englewood Cliffs, N.J.: Prentice-Hall, forthcoming. *P*

Principles of kinesics illustrated with photographs from everyday life, and sections on the political aspects of kinesics.

766 **Scheflen, Albert E.** *The Stream and Structure of Communicational Behavior: Context Analysis of a Psychotherapy Session.* Rev. ed. Bloomington, Ind.: Indiana University Press, forthcoming. *D T*

A greatly revised edition of a monograph published in 1965, this book is the impressive result of many years of film research and theoretical development. Several years' study of one film and the program of interaction deciphered through painstaking film analysis using what Scheflen calls the "context-analysis" procedure is described. A report of the individuals and research influencing this work, how the author turned to a behavioral systems approach and away from individual-centered psychological research, is presented. What follows is an exhaustive description of the communicative behavior—much of it kinesic or nonverbal—of a thirty-minute family therapy session with a mother and daughter and two therapists, Carl Whitaker and Thomas Malone. The lexical-kinesic behavior units and levels are defined and documented in terms of what Scheflen calls "positions" and "points," the behaviors delimiting them, and the contexts and regular sequences in which they occur. "Nonlanguage points" (either physical tasks, behavior that maintains orientation, or representational behaviors) and their function as signals, subgrouping behaviors establishing relatedness, dominance, etc., and "language and metacommunication points" and their kinesic aspects are described and illustrated. Later chapters describe how these behavioral units are integrated hierarchically; how their performance is regulated; and what the various relationships between members are, such as complimentarity. Performance styles and strategies of the members are described within a developmental and cultural context, along with discussion of the lexical-kinesic characteristics of schizophrenics.

767 **Schilder, Paul.** "The Organic Background of Obsessions and Compulsions." *American Journal of Psychiatry* 44 (1938):1397-1413.

Referring to seven cases of severe obsessional neurosis with marked compulsive features and "motor impulse disturbances," the author presents information from the psychiatric assessment and treatment as well as neurological examination that suggests underlying organic factors. The subjects' behavior is compared to the compulsive behavior of postencephaletic patients.

768 **Schilder, Paul.** *The Image and Appearance of the Human Body.* New York: International Universities Press, 1950. 353 pp. *D* [Latest ed., 1958]

In this classic treatise on body image and its physiological, psychological, and sociological origins, there is a section on the basis of action, expressive movement, and kinesthetic perception (pp. 50-70), as well as scattered references to the relation between body image and movement, kinesthesis, and postural reactions, formulated within a psychoanalytic framework.

769 **Schiller, Claire H.,** ed. and trans. *Instinctive Behavior: The Development of a Modern Concept.* New York: International Universities Press, 1957. 328 pp. *D P* [Latest ed., 1964]

Translations of important contributions to ethology, this book includes papers by Konrad Lorenz on the origins and functions of instinctive actions and motor patterns and "innate releasing mechanisms," and his critique of other theories about the nature of instinct. The studies of instinctive motor actions in the grey-lag goose by Lorenz and Nicholas Tinbergen and the feeding behavior of thrushes by Tinbergen and D. J. Kuenen are described. Also included is a chapter by Paul H. Schiller, entitled "Innate Motor Action as a Basis of Learning-Manipulative Patterns in the Chimpanzee."

770` **Schlosberg, Harold.** "The Description of Facial Expressions in Terms of Two Dimensions." *Journal of Experimental Psychology* 44 (1952):229-237. *D T*

In four experiments, subjects sorted and rated pictures on a scale from unpleasantness to pleasantness and on another from rejection to attention. Analysis showed that two different series of facial expressions could be distributed over an oval surface with these two dimensions as the axes.

771 **Schlosberg, Harold.** "Three Dimensions of Emotion." *Psychological Review* 61 (1954):81-88. *D*

Following a review of activation theory and its relation to emotion, Schlosberg presents preliminary evidence that the range of facial expressions (and perhaps emotions) can be described in terms of three dimensions: attention-rejection, pleasant-unpleasant, and sleep-tension. This formulation is schematically represented by an irregular cone with level of activation (sleep-tension) as the third dimension.

772 **Schmais, Claire.** "Dance as Part of the Pre-school and the Elementary School Curriculum." Unpublished paper, n.p., 1966. 11 pp. (Copies from author, Dance Therapy Masters Program, Hunter College, 695 Park Avenue, New York, N.Y.)

The author discusses the value of dance and music for the education of ghetto children and how dance may reinforce the children's cultural attributes. She outlines specific areas which can be developed through dance and role-playing, such as body image, language skills, and abstract concepts and discusses the value of training sessions using dance for teachers.

773 **Schneirla, T. C.** "An Evolutionary and Developmental Theory of Biphasic Processes Underlying Approach and Withdrawal." In *Nebraska Symposium on Motivation,* edited by M. R. Jones, pp. 1-42. Lincoln, Neb.: University of Nebraska Press, 1959. *D G P T*

An analysis of developmental and motivational characteristics of patterns of approach and withdrawal, considered the only "objective terms applicable to *all* motivated behavior in *all* animals" (p. 2). Hypothesizing that, in general, low-intensity stimulation evokes approach and high-intensity elicits withdrawal, the author traces approach and withdrawal movements from protozoa to mammals. This is extended into an analysis of emotional differentiation in man and the development of smiling and reaching in infants. An excellent bibliography is included.

774 **Schrier, Allan M.; Harlow, Harry F.; and Stollnitz, Fred.** *Behavior of Nonhuman Primates: Modern Research Trends.* Vol. 2. New York: Academic Press, 1965. pp. 287-595. *D G P T*

With the exception of the review of primate field studies by Phyllis Jay, the focus in this volume is on laboratory research of important areas, among them mother-infant interaction and "affectional systems" (Harry F. and Margaret K. Harlow), social behavior in young chimpanzees (William A. Mason), and changes with age in chimpanzees (A. J. Riopelle and C. M. Rogers).

775 **Schultz, Johannes H., and Luthe, Wolfgang.** *Autogenic Therapy.* Vol. I, *Autogenic Methods.* New York: Grune & Stratton, 1969. 255 pp. *G P T*

Originating from research and therapeutic use of hypnosis, autogenic therapy involves relaxation exercises and passive concentration on specific images and sensations of the body. In this volume the exercises are described, and trainees' reports of their experiences are reported in detail. Meditative exercises involving experience of colors and images and special "formulas" for treatment of specific physical and psychological symptoms are presented.

776 **Schutz, William C.** *Joy: Expanding Human Awareness.* New York: Grove Press, 1968. 223 pp.

An introduction to approaches being developed in the human-potential movement, this book includes chapters on psychological aspects of body structure and movement and therapeutic techniques using movement, touch, body awareness, and psychodrama. Throughout the book the author describes and gives the rationale for specific techniques or structured activities for encounter groups—many of these nonverbal—and presents examples of when they were used and how individuals reacted.

777 **Schwartz, Irving.** "Patterns of Communication in Families with Acting-Out Children as Compared to Families with Withdrawn Children." Ph.D. dissertation, New York University, 1967. 298 pp. (Datrix order no. 68-6103)

From judgments of acceptance or rejection in the verbal and nonverbal behavior of parents in filmed family therapy, it was found that the parents of aggressive children tended to express acceptance verbally and rejection nonverbally, whereas parents of withdrawn children expressed acceptance verbally with *both* nonverbal acceptance and rejection.

778 **Scott, Roland B.; Ferguson, Angella D.; Jenkins, Melvin E.; and Cutter, Fred F.** "Growth and Development of Negro Infants: V, Neuromuscular Patterns of Behavior during the First Year of Life." *Pediatrics* 16 (1955):24-30. *G T*

On the basis of reports of mothers and monthly tests by pediatricians, the development of specific motor patterns of black infants from two socioeconomic levels was compared with that of white infants. Between eight and fifty-two weeks of age, the black infants developed faster than the whites, as measured by types of head and hand control and the onset of rolling, sitting, crawling, pulling self up, walking with support, standing, and walking alone. Further, from eight to thirty-five weeks old, the "clinic" black infants progressed faster than the "private practice" black infants.

779 **Sebeok, Thomas A.** "Animal Communication." *Science* 147 (Jan.-Mar., 1965):1006-1014. *G P*

Discussion of the nature of animal communication and the problems of a new field, "zoosemiotics," in analyzing the communication code and classifying behavior. There are a number of references to movement as one channel of the system.

780 **Sebeok, Thomas A.,** ed. *Animal Communication: Techniques of Study and Results of Research.* Bloomington, Ind.: Indiana University Press, 1968. 686 pp. *D G P T*

Chapters on the nature of communication and problems of observation and analysis of animal behavior are followed by a cross-species comparison of visual signals and movements, notably aggressive, submissive, and sexual signals. There are eight chapters on communication systems of animals (including vocalizations, postures, displays, movements, etc.): those for arthropods, bees and other invertebrates, fishes, amphibians and reptiles, birds, land mamals, marine mammals, and primates. A chapter by A. Richard Diebold, Jr., extends this impressive survey to human communication, especially to research on mutual glancing and visual behavior. Of particular note is the paper on redundancy and coding in communication by Gregory Bateson.

781 **Sebeok, Thomas A.; Hayes, Alfred S.; and Bateson, Mary C.,** eds. *Approaches to Semiotics: Transactions of the Indiana University Conference on Paralinguistics and Kinesics.* The Hague: Mouton Publishers, 1964. 294 pp. *N T* [New York: Humanities Press]

This book contains two important papers on kinesics. The first, a chapter by Alfred S. Hayes entitled "Paralinguistics and Kinesics: Pedagogical Perspectives," is a clear summary of Birdwhistell's approach to kinesics research and contains proposals for kinesics and paralinguistics training for classroom teachers, linguists, and graduate students. The second, by Weston La Barre and entitled "Paralinguistics, Kinesics, and Cultural Anthropology," contains a valuable review. of literature, some of which is not included in the present bibliography, such as rare references to sign languages of different cultures. La Barre describes gestures of greeting, beckoning, contempt, etc., from around the world and cultural differences in gesture styles, both ancient and modern. He describes a fascinating potpourri of cultural movement conventions and motor behaviors little studied as yet. In addition to these chapters, there are interesting comments on kinesics by such participants as Margaret Mead, Ray L. Birdwhistell, Erving Goffman, Peter F. Ostwald, and Bernard Tervoort, including observations, definitions of terms, and theoretical remarks.

782 **Secord, Paul F.; Dukes, William F.; and Bevan, William.** "Personalities in Faces: I, An Experiment in Social Perceiving." *Genetic Psychology Monographs* 49 (1954):231-279. *G T*

The topic of this monograph is observer judgment of facial physiognomy, which includes fixed facial expressions and muscular tension as well as physical structure. Results of judgments of photographs show high agreement both on personality traits and physical descriptions, although subjects were hardpressed to explain their impressions of personality traits. Mouth curvature and facial tension were important determinants of specific traits, and several tables that are included indicate which physiognomic features were associated with which personality traits.

783 **Selden, Samuel.** *The Stage in Action.* Carbondale, Ill.: Southern Illinois University Press, 1941. 324 pp. *D P* [Latest ed., 1967]

This book on actor's training is unusual for its focus on dance and body movement in relation to acting. Particularly notable are the reviews of the emotive effects of historic, ethnic, and modern dance; the dramatic value of rhythm; "dance" and "mime" in an actual dramatic performance; the interpretation of expression in different postures, gaits, and gestures; analysis of action (as change, anticipation, or "transit"), and principles of group formation.

784 **Selver, Charlotte.** "Sensory Awareness and Total Functioning." *General Semantics Bulletin* 20-21 (1957):5-16. *P*

The author describes and illustrates an approach to sensory awareness and a freeing of body constrictions or inertia which she has developed in the tradition of the work of Elsa Gindler and Heinrich Jacoby. First, however, she describes the vital organic functioning of infants and how in time they become constricted by fear and conditioning.

785 **Shagass, Charles, and Malmo, Robert B.** "Psychodynamic Themes and Localized Muscular Tension during Psychotherapy." *Psychosomatic Medicine* 16 (1954):295-313. *G T*

EMG recordings during psychotherapy sessions were synchronized with recordings of the patient's speech. Three case studies of severely neurotic hospitalized patients are presented. Hostility themes were found to correspond with increased forearm tension; sex themes with leg tension in the female patients; and for one patient, high tension was associated with depressed mood. Methodology is extensively discussed.

786 **Shatan, Chaim.** "Unconscious Motor Behavior, Kinesthetic Awareness and Psychotherapy." *American Journal of Psychotherapy* 17 (1963):17-30.

Following a description of Gindler techniques for developing body awareness, the author discusses the role of fixed postures, tensions, and breathing restrictions in suppressing emotions and how some psychotherapists have used body-awareness techniques in psychotherapy. From observations in therapy and collaboration with a body-reorientation therapist, the author presents "kinesthetic-motor vignettes" about patients and interprets the significance of rigid, flaccid, and withdrawn motor states.

787 **Shawn, Ted.** *Every Little Movement: A Book about François Delsarte.* (M. Witmark & Sons, 1910; reprint ed., New York: Dance Horizons, 1963). 127 pp. *D*

During his life Delsarte (1811-1871), a musician who sought to formulate laws of art and expression, studied how people expressed themselves in all types of situations. This book presents his history, his theories, and their application to dance. Delsarte evolved a system for interpreting the psychological significance of various movement patterns in different parts of the body and different areas of space, and of specific postures and hand gestures. Included is an annotated bibliography of literature about Delsarte's work.

788 **Sheets, Maxine.** *The Phenomenology of Dance.* Madison, Wis.: University of Wisconsin Press, 1966. 158 pp.

A most interesting book on the dancer's and audience's experience of dance — in a sense, its "existential" analysis. There are chapters on the phenomenological analysis of dance as "virtual force" and "imaginative space," the nature of abstraction and expression in dance, and "dynamic line" and rhythm.

789 **Sheldon, W. H., in collaboration with S. S. Stevens.** *The Varieties of Temperament: A Psychology of Constitutional Differences.* New York: Harper & Brothers, 1942. 520 pp. *P T* [New York: Hafner Publishing Company, 1969]

A classic work on body physique and personality comparing Sheldon's three basic body types or mixtures thereof with temperament scales of types of "viscerotonia," "somatotonia," and "cerebrotonia." The traits are defined, and six case histories having numerous references to the individual's body movement patterns are presented. A study of adjustment and body type of 200 men is reported, followed by chapters on statistical analysis of the results and theoretical discussion. Data for the 200 cases are presented in the Appendix.

790   **Sherman, Mandel.** "The Differentiation of Emotional Responses in Infants: I, Judgments of Emotional Responses from Motion Picture Views and from Actual Observation." *Journal of Comparative Psychology* 7 (1927):265-284. *T*

Films or live observation of neonates in four situations (while hungry, after being dropped a small distance, after restraint of the head, after a slight needle prick) were judged by various groups, with some seeing only the behavior, some told the situation, and some told incorrect situations. There was very poor agreement among all groups, except for the untrained observers who knew the stimuli.

791   **Shimota, Helen E.** "The Relation of Psychomotor Performance to Clinical Status and Improvement in Schizophrenic Patients." Ph.D. dissertation, University of Minnesota, 1956. 139 pp. (Datrix order no. 18-949)

Normal and schizophrenic subjects were given a battery of structured tests of gross and fine motor skill. Performance distinguished the groups in a number of ways, and clinical improvement after treatment was related to improved performance on gross motor skills.

792   **Shirley, Mary M.** *The First Two Years: A Study of Twenty-Five Babies.* Vol. 1, *Postural and Locomotor Development.* Minneapolis: University of Minnesota Press, 1931. 227 pp. *G P T* [Westport, Conn.: Greenwood Publishing Corporation, 1971]

Based on extensive examination of the infants at regular intervals over two years, this book reports the stages of locomotor development, from neonate postures to assuming upright posture to walking. Age of onset, individual differences, the relation of locomotor development to muscle tone and physical growth, and its low correlation with intellectual ability are analyzed in detail. There is also a section on the development of locomotor play. The record sheets used for daily observation by the mothers are reprinted.

793   **Shirley, Mary M.** "Locomotor and Visual-Manual Functions in the First Two Years." In *A Handbook of Child Psychology,* 2d rev. ed., edited by C. Murchison, pp. 236-270. Worcester, Mass.: Clark University Press, 1933. *D G P T* [New York: Russell & Russell Publishers, 1967]

This valuable review of early research on motor development includes studies of fetal activity, newborn "mass activity" and "impulsive movements," early reflexes, and the sequence in which eyes, limbs, and total body develop control in the human child and in various animals. Rhythms and maturational patterns of development are discussed. Detailed studies of prehension and walking, as well as individual differences in rate and the relation between motor development and intelligence, are focused on.

794   **Shirley, Mary.** "A Behavior Syndrome Characterizing Prematurely Born Children." *Child Development* 10 (1939):115-128. *T*

Among the behaviors described as characterizing a group of two-month premature children and not full-term children who were research controls are retardation in development of prehension and locomotion, with peculiar gestures, awkward mobility, and either hyper- or hypoactivity. The author theorizes that the sensory capacities of these children are more developed than their motor control is.

795 **Shuey, Herbert.** "An Investigation of the Luria Technic with Normal and Psychotic Subjects." *Journal of Abnormal and Social Psychology* 32 (1937): 303-313. *T*

For this study an apparatus was developed to record the pattern of hand pressure when a subject responded to a stimulus word with his association and with a press of a lever. The author describes four motor patterns obtained: small (i.e., weak pressure), large, transition (both at different times), and "chaotic." He asserts that these patterns distinguished between normal and psychotic individuals and between different types of schizophrenics.

796 **Siegel, Marcia B.** "Maximiliano Zomosa (1937-1969)." *Dance Magazine* (Feb., 1969):29. *P*

A dance critic's description of the performance of the late dancer as Death in the Joost ballet *Green Table* a compelling glimpse of how the dancer communicates powerful emotions.

797 **Silver, Archie.** "Postural and Righting Responses in Children." *Journal of Pediatrics* 41 (1952):493-498.

A description of the postural reflex tests and case descriptions of an "organic" child and a very compulsive child with epilepsy are presented. Abnormal postural responses and levels of development in schizophrenic children are described.

798 **Silverman, Jacob J.** "Study of Tremor in Soldiers." *A.M.A Archives of Internal Medicine* 82 (1948):175-183. *P T*

Results of a study using an apparatus to measure tremor and a chemically treated paper to measure palmar sweat. High-amplitude tremor and excessive sweat characterized the group suffering from anxiety ("neurocirculatory asthenia").

799 **Simmel, Georg.** "Sociology of the Senses: Visual Interaction." In *Introduction to the Science of Sociology*, edited by R. E. Park and E. W. Burgess, pp. 356-361. Chicago: University of Chicago Press, 1921. [Latest ed., 1969]

An early paper on the social function of visual behavior and mutual glance and what may be communicated through the eyes.

800 **Singer, Jerome L., and Herman, Jack.** "Motor and Fantasy Correlates of Rorschach Human Movement Responses." *Journal of Consulting Pschology* 18 (1954):325-331. *T*

Schizophrenic subjects who had a high number of Rorschach M responses showed greater ability to inhibit (here measured as how slowly one could write), less activity while waiting, and signs of greater imaginative capacity compared with those having a low number of M responses. This is considered support for theories concerning a relation between motor control, kinesthetic perception on the Rorschach, and fantasy.

801 **Singer, Jerome L., and Opler, Marvin K.** "Contrasting Patterns of Fantasy and Motility in Irish and Italian Schizophrenics." *Journal of Abnormal and Social Psychology* 53 (1956):42-47. *T*

Patients were given a series of tests, including the Rorschach, the Porteus Maze, and the slow writing task of Downey's Will-Temperament Scale. The Irishmen showed more Rorschach M responses; were more cautious in their approach to the mazes; showed longer inhibition in the writing; and were less aggressive in their behavior in the hospital. The authors also note that the Irish patients were less likely to show catatonic excitement, and the Italians were less likely to show "flattened affect."

802   **Singleton, W. T.** "The Change of Movement Timing with Age." *British Journal of Psychology* 45 (1954):166-172. *D G*

Subjects from twenty to seventy years of age performed mechanical tasks that involved shifting levers in fixed directions as quickly as possible. It was found that the older men did not have slower movements over a distance but did spend more time at points involving a change of direction.

803   **Sloan, William.** "Motor Proficiency and Intelligence." *American Journal of Mental Deficiency* 55 (1951):394-406. *T*

Twenty retarded ten-year-olds and twenty normal children were given the "Lincoln Adaptation" of the Oseretsky Tests of Motor Proficiency and the Vineland Social Maturity Test. Subjects were matched for age and sex, and none showed organic pathology. The mentally retarded children scored significantly lower on all six parts of the Oseretsky tests, with their worst performance being on measurements of "simultaneous movement." Performances on the motor tests also correlated significantly with the level of social maturity as measured.

804   **Sloan, William.** "The Lincoln-Oseretsky Motor Development Scale." *Genetic Psychology Monographs* 51 (1955):183-252. *G P T*

The norms of the tests for children ages six to fourteen are presented in this monograph, which includes a description of the scale, its development, reliability and validity, administration, and scoring, illustrated with photographs of children performing specific motor tasks.

805   **Smith, A. Arthur.** "An Electromyographic Study of Tension in Interrupted and Completed Tasks." *Journal of Experimental Psychology* 46 (1953):32-36. *D G T*

Results show that muscle tension does not fall so greatly after an interrupted task as after a completed one.

806   **Sommer, Robert.** "Small Group Ecology." *Psychological Bulletin* 67 (1967): 145-152.

A review and bibliography of research on group formation and seating arrangements in face-to-face groups, describing their relation to status, "communication flow," competition, tolerance for closeness, cultural and sex differences, and so on.

807   **Sommer, Robert.** *Personal Space: The Behavioral Basis of Design.* Englewood Cliffs, N.J.: Prentice-Hall, 1969. 177 pp. *D G T*

Within this broad analysis of the interaction between physical environment, spacing, and behavior, studies of how people respond to "spatial invasion," the social significance of various seating arrangements (e.g., at round versus square tables), and the spatial behavior of patients in a mental hospital are reported, including observations of nonverbal behavior.

808   **Sorell, Walter,** ed. *The Dance Has Many Faces.* 2d rev. ed. New York: Columbia University Press, 1966. 276 pp. *P*

A collection of articles by noted dancers and choreographers, some of them dealing with aesthetic, psychological, or therapeutic aspects of dance.

809   **Sorell, Walter.** *The Dance through the Ages.* New York: Grosset & Dunlap, 1967. 304 pp. *D N P*

One of a number of broad histories of ethnic and theatrical dance, beautifully illustrated with photographs, this book is notable for sections on Roman and Greek rituals, the dances of the Far East (India, China, Korea, Japan, and the Pacific Islands), and the history of dance notations.

810 **Southwick, Charles H.** *Primate Social Behavior.* Princeton, N.J.: D. Van Nostrand Company, 1963. 191 pp. *D G P T*

A collection of papers on primate field and laboratory studies made over thirty years, including work by Zukerman, Carpenter, Haddow, Imanishi, and Kawamura.

811 **Speck, Frank G., and Broom, Leonary, in collaboration with Will West Long,** *Cherokee Dance and Drama.* Berkeley, Calif.: University of California Press, 1951. 106 pp. *D P*

Perhaps one of the most detailed records of American Indian dance and ritual, this work is the result of observations by ethnologists and of recall and reports of a member of the Cherokee. Some thirty-one dances and rituals are carefully described and the formations are illustrated. The "meanings" of the movements and rituals are explained.

812 **Spence, Donald P., and Feinberg, Carolyn.** "Forms of Defensive Looking: A Naturalistic Experiment." *Journal of Nervous and Mental Diseas* 145 (1967): 261-271. *T*

The subject's visual behavior was observed behind a one-way screen while he was waiting and had the opportunity to read a note about which negative traits he had. Those who on tests had shown a need to exhibit desirable traits avoided looking and drew very detailed figure drawings. Those who did not have this need carefully scrutinized the note. Types of visual avoidance (closing the eyes and inexact perception) and the behavior of the subjects are interpreted in terms of their defensive function.

813 **Spence, Lewis.** *Myth and Ritual in Dance, Game, and Rhyme.* London: C. A. Watts & Co., 1947. 202 pp. *D P*

A "genealogy" of folk dances, games, and rhymes, the book explores their origins in ancient religious rituals, myths, and sports. The function of rituals and dance in ancient Mayan, Greek, Roman, and Egyptian society and in various other cultures around the world, as well as the origins of certain European ceremonies, festivals, and superstitions, is analyzed.

814 **Sperry, Bessie; Ulrich, David N.; and Staver, Nancy.** "The Relation of Motility to Boys' Learning Problems." *American Journal of Orthopsychiatry* 28 (1958):640-646.

Clinical observations of a small group of preadolescent boys who showed disturbances in physical activity, such as severe fears of falling, accident proneness, and restlessness in the classroom. Put within a psychoanalytic framework, the paper consists largely of notes on their histories and their learning disabilities and specific case examples of how they use physical activity as a psychological defense.

815 **Spiegel, John P.** "Classification of Body Messages." *A.M.A. Archives of General Psychiatry* 17 (1967):298-305. *D N*

A system for defining and notating positions in terms of body spaces and the spatial relationship between subjects is presented, along with a summary of research done with Pavel Machotka on observers' interpretations of figures such as Boticelli's Venus in different poses.

816 **Spitz, René A.** "Hospitalism: An Inquiry into the Genesis of Psychiatric Conditions in Early Childhood." *Psychoanalytic Study of the Child* 1 (1945): 53-74. *G T*

Spitz's classic study of the devastating effects on infants of prolonged hospitalization in a foundling home, with little human contact as compared with the development of children raised at home or at a penal nursery under their mother's care. The foundling-home children showed much higher susceptibility to illness and had high mortality rates; a marked decrease in their "developmental quotient" after nine months; such restricted motility that they lay on their backs for months and played mainly with hands and feet; and in some cases bizarre, stereotyped movements.

817 **Spitz, René A.** "Hospitalism: A Follow-up Report on Investigation Described in Volume I, 1945." *Psychoanalytic Study of the Child* 2 (1946):113-117. *T*

From the group Spitz first studied, this follow-up includes twenty-one foundling-home children, between two and four years old and still institutionalized, although in better situations. Some of these children were unable to locomote in any way, and only five could walk unassisted. Most could not eat with a spoon or dress themselves.

818 **Spitz, René A.** *No and Yes: On the Genesis of Human Communication.* New York: International Universities Press, 1957. 170 pp. [Latest ed., 1966]

The author traces the development of head movements ("cephalogyric behavior") from the rooting response of the neonate to the head shake as a semantic signal for the two-year-old; and using this ontogenesis as illustration, he analyzes the development of the ego, object relations, and personality within a psychoanalytic framework. He includes chapters on animal infant movements, processes of imitation and identification, and the genesis of the head nod "yes."

819 **Spitz, René A., with the assistance of K. M. Wolf.** "The Smiling Response: A Contribution to the Ontogenesis of Social Relations." *Genetic Psychology Monographs* 34 (1946):57-125. *G P T*

A study of the emergence of the smiling response in infants and, by inference, differentiation of positive emotions and discrimination between subject and object: when smiling is first observed, what contexts evoke it in the first year, and what are its implications in social development. Studies in which faces, live or artificial, happy or stern, are presented indicate no racial or environmental differences; and that children, ages three to six months, smile at pleasant or ferocious faces and to face-front people or masks, not to details of the face. Different inanimate objects did not evoke the smiling response in children ages three to six months. Individual exceptions to these findings are illustrated with case descriptions, and evidence that mourning reactions occur in hospitalized infants over seven months old is interpreted.

820 **Spitz, René A., in collaboration with K. M. Wolf.** "Autoeroticism: Some Empirical Findings and Hypotheses on Three of Its Manifestations in the First Year of Life." *Psychoanalytic Study of the Child* 3-4 (1949):85-120. *G T*

A study of three autoerotic activities in infants: rocking, genital play, and fecal games in a group of white and black infants institutionalized in a penal nursery. The number of children showing each activity, and at what age, is presented. Racial differences are unclear. Rocking was most common in the nursery children, genital play in children raised at home. Preference for a specific activity (or lack of any) is related to the nature of the child's object relations, particularly the mother-child interaction.

821   **Spolin, Viola.** *Improvisation for the Theater: A Handbook of Teaching and Directing Techniques.* Evanston, Ill.: Northwestern University Press, 1963. 399 pp. *D P*

In this book on uses of games, improvisation, and group interaction in training the actor, particular emphasis is put on "physicalization" and freedom of physical, nonverbal expression. Numerous exercises and workshop procedures are described, many of which concentrate on body awareness, group movement, gaze behavior, gestures, facial expression, etc., and include examples with interpretations. These and chapters on expression of emotion and developing a character give an excellent, if rare, view of a director's analysis of movement.

822   **Stanislavski, Constantin.** *An Actor Prepares.* Translated by E. R. Hapgood. New York: Theatre Arts Books, 1936. 295 pp.

Chapters 3 and 6 ("Action" and "Relaxation of Muscles") are very interesting presentations of the role of movement in acting and actor training as conceived by Stanislavski. To illustrate such principles as the necessity of an inner justification, logic, and coherence for every action in order for there to be fluent, natural action and expression, the author describes acting sessions with considerable attention to movement and bodily expression.

823   **Stanislavski, Constantin.** *Building a Character.* Translated by E. R. Hapgood. New York: Theatre Arts Books, 1949. 292 pp.

Stanislavski presents his approach to acting in the form of an imaginary student describing the acting sessions of a master teacher who is really Stanislavski. So it is interesting, lively, and full of examples of encounters, incidents, and movement patterns observed. In this sequel to *An Actor Prepares*, there are a number of chapters explicitly on movement in acting, movement training for the actor, and how the actor creates "external" physical effects—characteristic facial expressions, postures, styles of walking and gesturing, tempo and rhythm—to portray a character. A fascinating example of how a great actor and director understands movement.

824   **Stauffacher, James C.** "The Effect of Induced Muscular Tension upon Various Phases of the Learning Process." *Journal of Experimental Psychology* 21 (1937): 26-46. *G T*

Three experiments showing that a certain degree of "induced" muscle tension facilitates learning and helps poor learners more than good learners.

825   **Stebbins, Genevieve.** *Delsarte System of Dramatic Expression.* New York: Edgar S. Werner, 1886. 271 pp. *D*

This presentation of Delsarte's "aesthetic science of expression" is very dated in writing style and conceptualization but interesting for its interpretations of areas of the body, body positions, styles of walking, hand gestures, arm movements, attitudes of the torso and head, and eye movements. Eye and facial expression is illustrated and interpreted in detail. "Laws of motion" and their relationship to emotion and intellect are presented. Chapters on the development of breathing habits and voice training are included. For a related in-print book by the author, see *Delsarte: Physical Training,* Hackensack, N.J.: Wehman Brothers, 1950.

826   **Stepanov, V. I.** *Alphabet of Movements of the Human Body.* Translated

by R. Lister (French ed., 1892). Cambridge: Golden Head Press, 1958. 47 pp. *D N* [Brooklyn, N.Y.: Dance Horizons, 1969]

A system of movement notation originally developed in Russia for recording dances. It uses the musical staff and notes, plus special signs for position and ad- or abduction, flexion or extension, and rotation of arms and legs, musical dynamic terms for movement dynamics, numbers for body turns, and so on.

827 **Stern, Edith M.** "She Breaks through Invisible Walls." *Mental Hygiene* 41 (1957):361-371

A somewhat popularized but vivid and interesting description of some dance therapy sessions of the late Marian Chace, one of the "founders" of dance therapy in the United States. A number of dramatic encounters with seriously ill psychiatric patients are reported to illustrate Miss Chace's style, her dance therapy techniques, and her interpretations of movement behavior. These reports and the brief history of her career are particularly valuable, because Marian Chace has influenced so many dance therapists in the United States.

828 **Strauss, Erwin W.** "Rheoscopic Studies of Expression Methodology of Approach." *American Journal of Psychiatry* 108 (1951-52):439-443.

Presents an outline of the rationale, practical procedures, and filming equipment to be used in a laboratory for the study of expressive motion at the Lexington (Kentucky) V.A. Hospital.

829 **Strodtbeck, Fred L., and Hook, L. Harman.** "The Social Dimensions of a Twelve-Man Jury Table." *Sociometry* 24 (1961):397-415. *D T*

A study of sixty-nine experimental deliberations indicated that where jurors sit is related to and may affect the selection of a foreman and preferences for individual jurors. An analysis of spatial distance and position factors in relation to preference is presented.

830 **Strongman, K. T., and Champness, B. G.** "Dominance Hierarchies and Conflict in Eye Contact." *Acta Psychologica* 28 (1968):376-386. *T*

The eye contact, directed gaze, and speech with gaze patterns of subjects getting acquainted were recorded and compared with their ratings of whom they like. A significant correlation was found between a greater amount of speech with gaze and mutual attraction; analysis of eye contact and gaze aversion patterns indicated a dominance hierarchy; and a method for determining the predominance of "approach forces" is described.

831 **Stutsman, Rachel.** *Mental Measurement of Preschool Children: With a Guide for the Administration of the Merrill-Palmer Scale of Mental Tests.* Yonkers-on-Hudson, N.Y.: World Book Company, 1931. 368 pp. *D P T*

In addition to a presentation of the Merrill-Palmer Scale, this book has a valuable review of mental and perceptual-motor tests for children. Many of these structured tests include motor development and coordination items.

832 **Sullivan, Harry S.** *The Interpersonal Theory of Psychiatry.* Edited by H. S. Perry and M. L. Gawel. New York: W. W. Norton & Company, 1953. 393 pp. [Latest ed., 1968]

Like Erving Goffman, Sullivan seems to have set the stage for the subsequent increase in group and family interaction research, with its attention to nonverbal aspects of interaction. Sullivan himself discusses facial and body expression only fleetingly (notably pp. 86-91, 145-148), but his analysis of how the infant learns facial expressions and develops the ability to discriminate the "forbidding gestures" of "bad mother" from the tone and movement of "good mother" are distinctive.

833   **Surwillo, Walter W.** "Psychological Factors in Muscle-Action Potentials: EMG Gradients." *Journal of Experimental Psychology* 52 (1956):263-272. G

This experiment yielded evidence that the slope or pattern of increase in EMG frequency can be increased with greater motivation.

834   **Swan, Carla.** "Individual Differences in the Facial Expressive Behavior of Preschool Children: A Study by the Time-Sampling Method." *Genetic Psychology Monographs* 20 (1938):557-650. N T

Twenty-five nursery children at the Yale Clinic of Child Development were observed three times for five minutes at monthly intervals for six months. Their facial and vocal expressive behavior was recorded with a notation system (e.g., A for attentive expression, T for talking, H for hand-to-face activity). The expressive behavior proved to be individually different, with indications that it was related to attendance, choice of play materials, personality characteristics, and so on. This monograph includes lists of which literature on children's social behavior includes study of facial expression, as well as a bibliography with a number of European papers on facial expression in children.

835   **Sylvester, Doris M.** "Three Studies in Hyperactivity: I, A Descriptive Definition of Hyperactivity." *Smith College Studies in Social Work* 4 (1963):2-26. T

A questionnaire for describing the behavior of children considered hyperactive is presented which includes sections on motor activity, speech, attention span, temperament, and personality traits. The motor section includes items on what the child does and the direction, range, tempo, and variation of his motor behavior in various situations. On the basis of the questionnaire, the behavior patterns of forty-one hyperactive children are described.

836   **Tagiuri, Renato, and Petrullo, Luigi,** eds. *Person Perception and Interpersonal Behavior.* Stanford, Calif.: Stanford University Press, 1958, 390 pp. G T

The major collection of research papers on person perception, among them a chapter by Paul F. Secord on the factors involved in judgment of photographs of facial expression. Because social perception involves three aspects integral to most movement research—subject, observer, and situation—many of the papers presented here, though not directly on judgment of body movement, are relevant. That is, the nature of inference processes and observer characteristics, the effects of various subject cues, and the role of the situation on the observer's judgment are often crucial issues in movement research.

837   **Takala, Martti.** "Studies of Psychomotor Personality Tests, I." *Annales Academie Scientiarum Fennicae Sarja-Ser. B Nide-Tom* 81, 2 (1953):1-130. G P T

A most valuable review and experimental evaluation of a number of structured motor tests as they correlate with aspects of personality, affect, or cognition. The tests, apparatus, measures derived, reliability, interpretation of results, and their intercorrelations with other motor and personality tests are presented. Tests focused on are the Expectancy Reaction Tests, the Luria Test, the Body Sway Test, and, especially, the Mira Myokinetic Psychodiagnostic Test. This book also contains an excellent review of research on body movement, muscle tension, and personality, including many articles not written in English.

838   **Taylor, J. E.; Potash, R. R.; and Head, Dorothy.** "Body Language in the Treatment of the Psychotic." In *Progress in Psychotherapy,* vol. 4, edited by

J. H. Masserman and J. L. Moreno, pp. 227-231. New York: Grune & Stratton, 1959.

Examples are given of how therapists' physical responses to the nonverbal behavior of regressed psychotic patients helped to bypass resistance and release affect.

839 **Tegg, William.** *Meetings and Greetings: The Salutations, Obeisances, and Courtesies of Nations.* London: William Tegg and Company, 1877. 312 pp. *D*

Descriptions of rituals and forms of greeting, particularly how royalty and religious authorities are addressed and greeted in different countries. The author describes greeting ceremonies depicted in the Bible, religious ceremonies, court presentations, and types of salutations around the world.

840 **Teicher, Joseph D.** "Preliminary Survey of Motility in Children." *Journal of Nervous and Mental Disease* 94 (1941):277-304. *P*

A study of postural reflexes and laterality in children, ages four to thirteen, with various psychiatric problems and a possible "organic factor" is reported, following a review of research on postural reflexes. A series of structured tasks are described. Healthy and divergent patterns for each age are described, and case examples are presented. The relations between motor pathology and organic damage, motility and I.Q., sex and racial differences, and laterality and reading disability are discussed.

841 **Teltscher, Herry O.** "A Study of the Relationship between the Perception of Movement on the Rorschach and Motoric Expression." Ph.D. dissertation, Yeshiva University, 1964. 107 pp. (Datrix order no. 65-4118)

Twenty-five college students considered motorically active because they were athletes showed significantly less Rorschach Movement (M) than twenty-five non-athletic students.

842 **Tervoort, Bernard T., S.J.** "Esoteric Symbolism in the Communication Behavior of Young Deaf Children." *American Annals of the Deaf* 106 (1961): 436-480. *D*

Drawing on film analysis of Dutch and American deaf children engaged in conversation, the author discusses the nature and origins of natural gestures, their development into formal signs, and the role of imitation, associative recognition, and context in these patterns.

843 **Thams, Paul F.** "A Factor Analysis of the Lincoln-Oseretsky Motor Development Scale." Ph.D. dissertation, University of Michigan, 1955. 193 pp. (Datrix order no. 11362)

A factor analysis of the scores on the Lincoln-Oseretsky Test for 211 boys from seven and one-half to twelve and one-half years old showed that, contrary to expectation, the test does not measure six distinct areas of motor proficiency; rather, there is evidence that it measures a general motor factor. But the author argues that extensive validation of the tests should be done before it is used as a diagnostic instrument.

844 **Thayer, Stephen.** "The Effect of Interpersonal Looking Duration on Dominance Judgments." *Journal of Social Psychology* 79 (1969):285-286.

In a study involving experimenters who looked directly at subjects for long periods or for very short periods, subjects judged those who looked longer as dominant and felt they themselves were judged as less dominant. However, there was no difference in the subjects' own visual behavior in either case.

845   **Thayer, Stephen, and Schiff, William.** "Stimulus Factors in Observer Judgment of Social Interaction: Facial Expression and Motion Pattern." *American Journal of Psychology* 82 (1969):73-85. *D T*

Extending the work of Heider and Simmel, and of A. Michotte on the importance of movement in judgment of social interaction, the authors made short films of schematic "happy" or "sad" faces variously approaching or moving away from each other. Both the facial expression and the motion influenced observer judgment of hostile or friendly encounters, with the two stimuli interacting in a complex way.

846   **Thie, Joseph A.** *Rhythm and Dance Mathematics.* Minneapolis, Minn.: Joseph A. Thie, 1964. 101 pp. *D T*

A mathematical analysis of dance rhythms, their degree of "randomness," digital analysis, and the synthesis of complex rhythms. Also described are preliminary forays into computer-programming dance rhythms and notations of dance and use of the computer in choreography. The examples are relatively simple; this is a work written by a mathematician but designed to show performing artists the possibilities of the computer and mathematics for dance.

847   **Thompson, Diana Frumkes, and Meltzer, Leo.** "Communication of Emotional Intent by Facial Expression." *Journal of Abnormal and Social Psychology* 68 (1964):129-135.

An experiment in the ability of individuals to express ten specific emotions and the corresponding agreement in judging their facial expressions. There were individual differences in expressive ability unrelated to personality as assessed with the CPI. Recognition was better than chance, especially for judging expressions of "happiness, love, fear and determination."

848   **Thompson, Jane.** "Development of Facial Expression of Emotion in Blind and Seeing Children." *Archives of Psychology*, no. 264, 37 (1940):1-47. *G P T*

Twenty-six blind children and twenty-nine sighted children, ages seven weeks to thirteen years, were filmed in everyday situations and in situations designed to elicit certain expressions. Analysis of the results showed, for example, that facial activity in laughing, smiling, and crying was similar in both groups; that the amount of activity in smiling decreases with age in the blind children, especially those blind from birth; and that expressions of both groups were judged equally well. The author discusses the results as evidence of the innate nature of facial expression.

849   **Thornton, G. R.** "The Effect upon Judgments of Personality Traits of Varying a Single Factor in a Photograph." *Journal of Social Psychology* 18 (1943):127-148. *T*

Photographs were made of the same subjects, with or without glasses, with or without a smile, or with a dark or light print. Judges rated the subjects with glasses and a serious expression as more intelligent and smiling subjects as more honest and kindly.

850   **Thornton, S.** *A Movement Perspective of Rudolf Laban.* London: Macdonald & Evans, forthcoming.

A prepublication announcement stated that this is a book about Rudolf Laban, his life, his theories about movement analysis, and his influence on movement training in education.

851   **Tinbergen, N.** *Social Behaviour in Animals: With Special Reference to Vertebrates.* London: Methuen & Co., 1953. 150 pp. *D G P*

The mating behavior, family and group life, territorial and aggressive behavior, and social organization of a variety of birds, insects, and fish. Their "dances" are carefully described in this work on the nature and function of animal social behavior.

852   **Tinklepaugh, O. L., and Hartman, Carl G.** "Behavior and Maternal Care of the Newborn Monkey (Macaca Mulatta—'M. Rhesus')." *Journal of Genetic Psychology* 40 (1932):256-86. *P T*

Logs of neonate activity and maternal reactions of several captive monkeys and description of sensorimotor development of the infants.

853   **Tobach, Ethel,** ed. "Experimental Approaches to the Study of Emotional Behavior." *Annals of the New York Academy of Sciences* 159 (1969):621-1121. *D G P T*

The following research reports from this important conference are notable for analysis of body expression or activity patterns: "Analyzing the Roles of the Partners in a Behavioral Interaction—Mother-Infant Relations in Rhesus Macaques" by R. A. Hinde; "Effects of Separation from Mother on the Emotional Behavior of Infant Monkeys" by I. Charles Kaufman and Leonard A. Rosenblum; "The Effects of Severe Restriction on Infant Dogs" by Ronald Melzak; "The Effects of Electrical Stimulation to the Brains of Monkeys" by José M. R. Delgado and Diego Mir; and " 'Open-field' Tests of Monkeys, Cats and Rats" by Douglas K. Candland, Z. Michael Nagy, and Victor H. Denenberg.

854   **Todd, Mabel Elsworth.** *The Thinking Body: A Study of the Balancing Forces of Dynamic Man.* (New York: Paul B. Hoeber, 1937; reprint ed., New York: Dance Horizons, 1968). 314 pp. *D*

The analysis of upright posture and locomotion in relation to principles of mechanics, anatomy, neurophysiology, and breathing. Treatment and corrective work utilizing body awareness, kinesthetic perception, and imagery is described.

855   **Tomkins, Silvan S.** *Affect Imagery Consciousness.* Vol. 1, *The Positive Affects.* New York: Springer Publishing Co., 1962. 522 pp. *D*

This monumental work on the nature of affect has several chapters which deal directly with movement and bodily expression. Of particular note is the chapter on the face as the "primary site of the affects," recognition of emotion from facial expression, facial expression styles, and physiological analysis of facial expression. There are also sections on the determinants of affect, including "affect motor passages," and facial and bodily expressions of interest-excitement, joy, and surprise-startle. Many aspects of emotion and its nature, function, and development are explored in the light of a great deal of research literature.

856   **Tomkins, Silvan S.** *Affect Imagery Consciousness.* Vol. 2, *The Negative Affects.* New York: Springer Publishing Co., 1963, 580 pp. *D G T*

Tomkins continues to analyze the facial and body expressions of emotion within his treatise on the nature of affect, focusing in this volume on distress-anguish, shame-humilation, and contempt-disgust. He develops theories about the forms and significance of emotional expression in children and their modification with socialization, as well as a theory of personality strongly based on an analysis of affect.

857   **Tomkins, Silvan S., and McCarter, Robert.** "What and Where Are the Pri-

mary Affects? Some Evidence for a Theory." *Perceptual and Motor Skills* 18 (1964): 119-158. *G T*

The authors present theories about the origin and nature of facial expressions and argue that the face is the organ for "maximal transmission of information" about affect to self and others and that there are eight primary affects. In their research they presented sixty-nine photographs of faces showing "neutral feeling" or one of the primary affects to judges who rated them according to words within each affect category. Subjects accurately named the primary affects; commonly confused certain terms for the primary ones; and showed individual biases. An extensive discussion about the relation between affects is presented.

858 **Tomkins, William.** *Indian Sign Language.* (San Diego, Calif., 1931; reprint ed., New York: Dover Publications, 1969). 106 pp. *D P*

Originally entitled *Universal Indian Sign Language of the Plains Indians of North America,* this book includes a dictionary of words and how they are indicated in sign language. The signs are illustrated with drawings that reflect the correct motion, and French and German equivalent words are cited. Sentence formation is illustrated, and samples of pictography are presented. Especially used by the Boy Scouts, the book has sections on sign language for scout initiation ceremonies.

859 **Traisman, Alfred S., and Traisman, Howard S.** "Thumb- and Finger-Sucking: A Study of 2,650 Infants and Children." *Journal of Pediatrics* 52 (1958):566-572. *T*

Following a brief review of papers on the subject, the authors report a longitudinal study showing that 45 percent of the children were thumb-suckers in the first year, that these children tended to feed longer, and that no relationship was found between thumb-sucking and psychological problems, colic, or sex differences.

860 **Tredgold, A. F.** *A Text-book of Mental Deficiency (Amentia).* (1908; reprint ed., Baltimore: The Williams & Wilkins Company, 1952). 545 pp. *P T*

This contains a brief section on body movement characteristics of mental defectives (pp. 135-137), with particular note of abnormal movements, "incoordination," and inertia or motor restlessness.

861 **Triandis, Harry C., and Lambert, William W.** "A Restatement and Test of Schlosberg's Theory of Emotion with Two Kinds of Subjects from Greece." *Journal of Abnormal and Social Psychology* 56 (1958):321-328. *D T*

Fifteen people from a Greek village and fifteen students from Athens were asked to rank photographs of facial expressions according to Schlosberg's dimensions for emotions (pleasant to unpleasant, attention to rejection, and sleep to tension). Both groups did so similarly to the judgments of American subjects, but the village group showed some interesting differences. The authors analyze Schlosberg's theory and propose a modification of the "solid" that illustrates the relationship between the dimensions.

862 **Tricker, R. A. R., and Tricker, B. J. K.** *The Science of Movement.* New York: American Elsevier Publishing Company, 1967. 284 pp. *D G P T*

This comprehensive analysis of the mechanics of human motion on land, in water, and when weightless also includes sections on the horse's gait and on bird flight.

863 **Trussell, Ella M.** "The Relation of Performance of Selected Physical Skills

to Perceptual Aspects of Reading Readiness in Elementary School Children." Ph.D. dissertation, University of California, Berkeley, 1966. 64 pp. (Datrix order no. 67-8504)

Seventy-five first and second graders were given the Frostig Developmental Test of Visual Perception, the Lincoln-Oseretsky Motor Development Scale, and eye and hand dominance tests; these were compared with reading achievement tests. Although there were intercorrelations, none was good enough to predict performance on another, and the study casts doubt on the feasibility of using perceptual-motor assessments or training for children with reading problems.

864   **Turkewitz, Gerald; Gordon, Edmund W.; and Birch, Herbert G.** "Head Turning in the Human Neonate: Spontaneous Patterns." *Journal of Genetic Psychology* 107 (1965):143-158. *D T*

Observation of the head movements of twenty healthy, awake neonates in terms of frequency and degree of turning yielded three basic types: those who crossed the body midline, those with an arc of ninety degrees, and those with a range of forty-five degrees or less. Eighteen babies kept their heads predominantly to the right. Number of head movements was correlated with arm movements and number of times the hand touched the face. Head movements did not correlate with leg activity, sex differences, Apgar scores, or racial differences.

865   **Uhrbrand, L., and Faurbye, A.** "Reversible and Irreversible Dyskinesia after Treatment with Perphenazine, Chlorpromazine, Reserpine and Electroconvulsive Therapy." *Psychopharmacologia* 1 (1960):408-418. *P T*

Among 500 female psychiatric patients, 33 had a dyskinesia of mouth and tongue due to ECT or drugs, primarily Trilafon and chlorpromazine. In a number of cases, the abnormal movements did not stop after cessation of the drug. The symptom occurs most in older patients who have had prolonged treatment and in those with organic brain lesions.

866   **Uklonskaya, R.; Puri, Basant; Choudhuri, N.; Dang, Luthura; and Kumari, Raj.** "Development of Static and Psychomotor Functions of Infants in the First Year of Life in New Delhi." *Indian Journal of Child Health* 9 (1960):596-601. *G*

The motor development of 1,000 Indian infants from age one month to a year was examined. The age at which the children first performed each of the selected activities (smiling, gaze fixation, holding of the head, supporting upper trunk, holding objects, turning over, sitting, crawling, standing, toddling, and speaking) is reported.

867   **Van Hooff, J. A. R. A. M.** "Facial Expressions in Higher Primates." *Symposia of the Zoological Society of London* 8 (1962):97-125. *D P*

The expressions of monkeys and apes were observed in terms of a number of "expression elements" (variations in visual, facial, posture, and vocalization behavior). The author describes types of facial expressions (relaxed, alert, attack, etc.) and when they occur. He groups them into two categories, as associated with agonistic situations or with conflict between attraction and a tendency to flee.

868   **Van Hooff, J. A. R. A. M.** "A Component Analysis of the Structure of the Social Behaviour of a Semi-Captive Chimpanzee Group." *Experientia* 26 (1970): 549-550. *T*

Seven "components" accounted for most of the variance in chimpanzee behaviors observed over 200 hours: the "affinitive" or positive social system, the "play" system, the "aggressive" system, the "submissive" system, the "excitement" system, the "show" system, and the "groom" system, in that order.

869 **Van Iersal, J. J., and Bol, A. C. Angela.** "Preening of Two Tern Species: A Study on Displacement Activities." *Behaviour* 13 (1958):1-88. *D G*

A very detailed, well-documented account of the tern bird's preening behavior (bill preening, headshaking, head rubbing, etc.), when it occurs, how it varies, when it is a "displacement behavior," and how it may be conceptually understood. There is also a bibliography of other studies of the social behavior of birds.

870 **Van Lawick-Goodall, Jane.** "A Preliminary Report on Expressive Movements and Communication in the Gombe Stream Chimpanzee." In *Primates: Studies in Adaptation and Variability,* edited by P. C. Jay, pp. 313-374. New York: Holt, Rinehart and Winston, 1968. *G P T*

This is a well-illustrated account of the chimpanzees' facial expressions and the situations which elicit them; developmental stages of motor behavior; mother-infant interaction; frustration, submission, and flight behaviors; and aggressive movements, hand and arm "gestures," and expressive movements in sexual contexts.

871 **Van Vlack, Jacques.** "Filming Psychotherapy from the Viewpoint of a Research Cinematographer." In *Methods of Research in Psychotherapy*, edited by L. A. Gottschalk and A. H. Auerbach, pp. 15-24. New York: Appleton-Century-Crofts, 1966. *D*

A description of the equipment, studio setting, processing, and professional issues involved in making documentary research films of psychotherapy sessions at Eastern Pennsylvania Psychiatric Institute.

872 **Vinacke, W. Edgar.** "The Judgment of Facial Expressions by Three National-Racial Groups in Hawaii: I, Caucasian Faces." *Journal of Personality* 17 (1949):407-429. *T*

Japanese, Chinese, and Caucasian men and women judged photographs and checked a list of thirty emotions. Marked racial differences in recognition were not found, but the women showed greater agreement among them than the men.

873 **Vine, Ian.** "Communication by Facial-Visual Signals." In *Social Behaviour in Birds and Mammals,* edited by J. H. Crook, pp. 279-354. New York: Academic Press, 1970. *D P*

An excellent review of literature on head movement, facial expression and visual behavior in animals and man, with focus on theoretical questions about their signal properties, origin, and function. The chapter is marked by the author's distinctive mode of integrating past research and conceptualizing and interpreting the issues. Literature on head movement and facial signaling of animals below primates, particularly birds, is reported, followed by that of monkeys and apes. However, most of the review deals with humans: expression and recognition of affect from facial expression, facial-visual signals in "social perception" and in interaction (particularly reviewing the work of Kendon, Argyle, and Birdwhistell), and finally, developmental aspects of facial and visual behavior. Vine concludes with some theoretical formulations regarding the development and function of facial-visual signals.

874 **Wachtel, Paul L.** "An Approach to the Study of Body Language in Psychotherapy." *Psychotherapy: Theory, Research and Practice* 4 (1967):97-100.

168 UNDERSTANDING BODY MOVEMENT

From repeated viewing of a videotape, the author reports clinical observations of the body motion of a paranoid woman as it correlated with her speech and reflected her intrapsychic conflicts and psychological defenses.

875 **Wagner, Isabelle F.** "The Body Jerk of the Neonate." *Journal of Genetic Psychology* 52 (1938):65-77. *T*

Body jerks and startle responses of 197 neonates to various stimuli were recorded and analyzed in terms of the body parts moved and the variability of the response, its relation to general stirs, and its decrease over time.

876 **Wagoner, Lovisa C., and Armstrong, Edna M.** "The Motor Control of Children as Involved in the Dressing Process." *Journal of Genetic Psychology* 35 (1928):84-97. *T*

Tests of the ability of nursery school children to button and unbutton a jacket appeared to correlate with motor-performance tests and personality traits of self-reliance, perseverance, and interest in detail better than with intelligence. Age and sex differences are noted, and how individual children performed the activity is related to their personality and I.Q. scores.

877 **Walker, Kathrine Sorley.** *Eyes on Mime: Language without Speech.* New York: The John Day Company, 1969. 190 pp. *P*

This review of mime and dance mime around the world includes a chapter on sign language in daily life. There are scattered descriptions of body movements and their meaning.

878 **Wardwell, Elinor.** "Children's Reactions to Being Watched during Success and Failure." Ph.D. dissertation, Cornell University, 1960.

This dissertation could not be found in *Dissertation Abstracts* and apparently is not available from University Microfilms. However, because it was cited as a valuable work a number of times in literature on gaze behavior and social perception, it is listed here.

879 **Warner, Francis.** "Muscular Movements in Man, and Their Evolution in the Infant." *Journal of Mental Science* 35 (1889):23-44. *G*

Subtitled "A Study of Movement in Man, and Its Evolution, Together with Inferences as to the Properties of Nerve-Centres and Their Modes of Action in Expressing Thought," this article is historically interesting for the descriptions of eye movement, the startle response, developmental stages (including graphs of infant hand movements), and the relationship between body movement—or what the author calls "microkinesis"—and thought.

880 **Washburn, Margaret Floy.** *Movement and Mental Imagery: Outlines of a Motor Theory of the Complexer Mental Processes.* Boston:Houghton Mifflin Company, 1916, 231 pp.

The author explores and theorizes about the relationship between physical movement and learning, consciousness, mental imagery, memory associations, and imageless experiences.

881 **Washburn, Ruth Wendell.** "A Study of the Smiling and Laughing of Infants in the First Year of Life."*Genetic Psychology Monographs* 6 (1929):397-537. *P T*

This monograph begins with an excellent review of literature on smiling and laughing in adults and children. There follows a study of the smiling, laughing, frowning, and crying behavior of fifteen infants, ages eight to fifty-two weeks, observed for fifteen minutes each month. An observational checksheet of face, eye, and body movement is included. The monograph reports a great deal of information on developmental stages of the expressive behavior; types of individual differences; effects of various stimuli; and the distinction between form, function, and development of smiling versus laughing.

882 **Watson, John B.** "A Schematic Outline of the Emotions." *Psychological Review* 26 (1919):165-196.

In an article defining emotion as "an hereditary pattern-reaction involving profound changes of the bodily mechanism" (p. 165), movement and bodily expression of rage, fear, and love are among the changes described.

883 **Watson, John B.** *Behaviorism.* New York: People's Institute Publishing Company, 1924. 251 pp. *D T* [New York: W. W. Norton & Company, 1970]

The observations and stress on actual visible behavior of behavioristic psychology in Watson's time are closely akin to much current movement research, although notions of instinct, habit, and conditioning may not be so prominent. In this series of lectures, Watson includes a section on behaviors observed in neonates and the development of "human action systems" to age five; a behavioristic analysis of emotions and bodily expression; and conditioning of emotional reactions.

884 **Watson, O. Michael, and Graves, Theodore D.** "Quantitative Research in Proxemic Behavior." *American Anthropologist* 68 (1966):971-985. *D T*

This controlled study of thirty-two Arab and American students empirically confirms Hall's assumptions that Arabs stand closer, face more directly, touch more, have more direct visual contact, and speak louder than Americans.

885 **Watzlawick, Paul; Beavin, Janet H.; and Jackson, Don D.** *Pragmatics of Human Communication: A Study of Interactional Patterns, Pathologies, and Paradoxes.* New York: W. W. Norton & Company, 1967. 296 pp. *P*

"Pragmatics" refers here to the behavioral effects of communication and includes body language, but there are few descriptive examples of movement in the book. Of note here are the distinction between "digital" and "analogic" communication and the parallels between analogic and nonverbal communication (pp. 62-63).

886 **Webb, Warren W.; Matheny, Adam; and Larson, Glenn.** "Eye Movements as a Paradigm of Approach and Avoidance Behavior." *Perceptual and Motor Skills* 16 (1963):341-347. *G*

In a series of experiments subjects were shown four pictures arranged horizontally, one or two of which were either positive or negative. As they looked at the pictures their eye movements were filmed. Schizophrenic subjects showed different patterns than normals. In situations where a picture was previously associated with conflict or shock, subjects generated an "avoidance gradient" to the associated pictures.

887 **Weber, Marylou Adam.** "The Motor Behavior Characteristics of Children with Operant Language Disorder." Ph.D. dissertation, University of Arizona, 1966. 119 pp. (Datrix order no. 67-1075)

The motor development and coordination of eleven children, ages four to eleven, who had language disorders according to results on the Illinois Test of Psycholinguistic Ability were tested with the Lincoln-Oseretsky Motor Development Scale and a measure of laterality and dominance. A significant relationship was found between the degree of motor maturation and the degree of language maturation. These children showed great difficulty with balance, agility, handling of objects, etc., and a notable lack of laterality and dominance pattern.

888   **Weinman, Bernard S.** "Changes in Fine-Motor Performance Induced by Drug Therapy and a High Activity Ward Program: An Investigation of the Effects of a Tranquilizing Drug, Promazine Hydrochloride, upon the Fine-Motor Functions of Hospitalized Male Chronic Paranoid Schizophrenics under a High and Low Activity Ward Program." Ph.D. dissertation, New York University, 1958. 146 pp. (Datrix order no. 58-7626)

Comparison of the fine motor performance of four groups of chronic schizophrenics (differing regarding placebo and high-activity ward program, promazine hydrochloride and high-activity program, placebo and low-activity program, promazine hydrochloride and low activity) before the treatment period and six weeks later showed a significant difference in motor performance on only one of six motor tests, the Purdue Pegboard Pin-Placing Test. This improvement appeared only in the placebo-high-activity group, suggesting that the drug does not enhance the fine motor performance of these patients.

889   **Weiss, Paul.** "The Social Character of Gestures." *Philosophical Review* 52 (1943):182-186.

A critique of Mead's concept of "significant gestures" and his assertion that gestures are defined by the social responses to them.

890   **[Wells, Samuel B. ?]** *How to Read Character: A New Illustrated Hand-Book of Phrenology and Physiognomy, for Students and Examiners; with a Descriptive Chart.* New York: Fowler and Wells Company, 1890. 192 pp. *D*

Phrenology "consists in judging from the head itself, and from the body in connection with the head, what are the natural tendencies and capabilities of the individual." This historical curiosity shows how complex and in keeping with the cultural and philosophical prejudices of the time physiognomy readings really were. From the size of thirty-seven small regions of the skull, one could interpret everything from "amativeness" (Aaron Burr's profile and concomitant sexual promiscuity are cited as an example) to "sublimity" (which may combine with "veneration" or "ideality").

891   **Wenger, M. A.** "Some Relationships between Muscular Processes and Personality and Their Factorial Analysis." *Child Development* 9 (1938):261-276. *T*

Drawing on Pavlov's observations of individual differences in his animals' temperaments, the author proposes several hypotheses about muscle tension, personality, and emotion. A factor analysis of muscle-tension patterns, observations of activity, response-rate measures, and personality ratings of six institutionalized boys supports his hypotheses.

892   **Wenger, M. A.** "An Attempt to Appraise Individual Differences in Level of Muscular Tension." *Journal of Experimental Psychology* 32 (1943):213-225. *T*

Seventy-four children, ages six to twelve, were rated for "characteristic level of muscular relaxation" by people who worked with them. Reliability among the raters was good, and further factor analysis with physiological measures indicated the validity of the relaxation ratings. Correlations of the ratings with personality characteristics are presented.

893   **Werner, Heinz.** "Motion and Motion Perception: A Study on Vicarious Functioning." *Journal of Psychology* 19 (1945):317-327.

Werner integrates research of his own showing that brain-injured children who are hyperactive show little perception of motion (whether stroboscopic, illusory, or empathic as in Rorschach M responses) and some fascinating German studies of the relationship between motor activity or eye movements and visual or kinesthetic perception of motion. He proposes that there is a dynamic property common to both sensory and motor processes—called variously tonicity, tonic innervation, or tonic energy—which may be released through body movement or sensory processes. In this formulation, perceptual space involves several perceptual modes in a "sensori-tonic field."

894 **Werner, Heinz.** *Comparative Psychology of Mental Development.* New York: John Wiley & Sons (Science Editions), 1948. 564 pp. *D T* [rev. ed., New York: International Universities Press, 1966]

Within this classic work on the comparison of the mental development of children, primitives, and psychotics, there are sections on the nature of primitive motor activity and action. Applying some of his key concepts of mental functioning to motor behavior, Werner discusses the intimate relationship between somatic-motor activity and primitive emotional experience, action as bound to concrete situation in primitive behavior, and uncoordinated, mass activity and rigidity as characteristic of primitive motor activity. Werner's exposition is extensively supported by references to empirical studies.

895 **Westcott, Roger W.** "Introducing Coenetics: A Biosocial Analysis of Communication." *American Scholar* 35 (1965-66):342-356. *T*

Westcott proposes the term "coenetics" to refer to the study of "interagent contact" of living organisms (with kinesics being one of the communication systems) and discusses the evolution of biosocial communication.

896 **Weston, Donald L.** "Motor Activity and Depression in Juvenile Delinquents." Ph.D. dissertation, Boston University, 1958. 138 pp. (Datrix order no. 58-3128)

Investigation of the theory that, for the juvenile delinquent in a situation of stress, there is an increase in depression against which he defends with increased motor activity. Thirty boys between twelve and sixteen years old were given Rorschachs (depression was measured by six scores from these) and the Katona matchstick problems to determine the increase in motor activity just before and then after their court appearances. According to the results, there was a significant increase in depression and in motor activity just before the adolescents were to appear in court.

897 **Whatmore, George B., and Ellis, Richard M.** "Some Neurophysiologic Aspects of Depressed States: An Electromyographic Study." *A.M.A. Archives of General Psychiatry* 1 (1959):70-80. *G T*

EMG recordings of residual motor activity during a relaxed position in severely depressed patients with motor retardation, depressed patients without motor retardation, and normal controls were made. Residual motor activity for both depressed groups was significantly higher than in the control group, with jaw-tongue activity being the greatest.

898 **Whiffen, Thomas.** *The North-West Amazons.* New York: Duffield and Company, 1915. 319 pp. *P*

A chapter on the dances of these Indians and notes on the meaning of certain gestures are included.

899 **White, Burton L.; Castle, Peter; and Held, Richard.** "Observations on the

Development of Visually-Directed Reaching." *Child Development* 35 (1964):349-364. *P T*

The visual-prehensile response to an object of thirty-four normal infants was studied under standardized conditions, and a normative sequence of eight stages was determined from one to five months of age, when visually directed reaching was established.

900   **White, Edwin C., and Battye, Marguerite.** *Acting and Stage Movement.* New York: Arc Books, 1963. 182 pp. *D P*

The second section of this book is devoted to movement: how the actor may sit, walk, hold his hands, move his head, arms, eyes, etc., and orient himself to others and express emotion. Particular attention given to sex and age differences in movement and positions is interesting sociological observation in its own right. The second author proposes movement exercises for the actor and an "Emotional Scale of Relaxation to Tension," with different emotions corresponding to different degrees of tension.

901   **White, Elissa Q.** "Child Development Movement Studies (Part I)." Unpublished paper, Goddard College, 1968. 48 pp. *T* (Copies from Dance Notation Bureau, 8 East 12th St., New York)

A film study of the movement characteristics of sixteen infants, eight to thirty-two weeks old, using a checklist based on Laban's effort-shape analysis. Age differences in amount of upper- and lower-body activity and in the complexity of the movement were found.

902   **Wickes, Thomas A., Jr.** "Examiner Influence in a Testing Situation." *Journal of Consulting Psychology* 20 (1956):23-26. *T*

Smiling, nodding, or moving forward after a subject gave an M response on the Rorschach increased the M responses significantly, as did saying "good" or "fine."

903   **Wigman, Mary.** *The Language of Dance.* Translated by W. Sorell. Middletown, Conn.: Wesleyan University Press, 1966. 118 pp. *D P*

A very noted dancer and choreographer looks back on a fifty-year career and discusses dance aesthetics, the process of creating dances, the interpretation of expression, and the nature of dance. Includes many photographs of her works.

904   **Williams, Frederick, and Tolch, John.** "Communication by Facial Expression." *Journal of Communication* 15 (1965):17-27. *P T*

Factor analysis of judgments of facial expressions yielded two dimensions: general evaluation and dynamism. When actors and nonactor subjects "encoded" them and made facial expressions for the dimension terms, judgments of their expressions corroborated the two-dimensional model. The actors were no more effective than the nonactors in making the expressions. Also, individuals tended to have different "neutral" expressions, which themselves were judged as more or less pleasant.

905   **Williams, Judith R., and Scott, Roland B.** "Growth and Development of Negro Infants: IV, Motor Development and its Relationship to Child Rearing Practices in Two Groups of Negro Infants." *Child Development* 24 (1953):103-121. *T*

The gross motor items of the Gesell Developmental Schedules were administered to 104 Negro infants of two different socioeconomic groups, and observations of child-raising practices, particularly in regard to handling and freedom of movement, were made in the home. The infants of the lower class showed greater acceleration of motor development, but the authors present evidence that faster motor development in Negro infants is not a function of racial differences but of a more accepting and permissive environment and less motor restriction.

906   **Winick, Charles, and Holt, Herbert.** "Seating Position as Nonverbal Communication in Group Analysis." *Psychiatry* 24 (1961):171-182.

Drawing on twenty years of experience in group psychotherapy and repeated experimentation with various seating arrangements, chairs, and rooms, the authors describe how patients typically behave in relation to the seating arrangement, what their preferences reflect, how slow group members are to initiate change in this, what a given chair may symbolize, and so on.

907   **Winick, Charles, and Holt, Herbert.** "Eye and Face Movements as Nonverbal Communication in Group Psychotherapy." *Journal of the Hillside Hospital* 11 (1962):67-79.

A very interesting discussion of patterns of gaze behavior, facial expression, and body movements observed in group therapy sessions with description of various types of gaze behavior and their significance for the individual or group; how the presence of a blind member changed the communication patterns of the group; and what specific mouth mannerisms meant to certain members.

908   **Witkin, H. A.; Dyk, R. B.; Faterson, H. F.; Goodenough, D. R.; and Karp, S. A.** *Psychological Differentiation: Studies in Development.* New York: John Wiley & Sons, 1962. 418 pp. *P T*

Within the extensive research on "field dependence" and personality reported here are a few studies on the correlation between field independence-dependence scores and gaze behavior, posture, and amount of activity in children.

909   **Wolff, Charlotte.** *A Psychology of Gesture.* Translated from the French by A. Tennant. London: Methuen & Co., 1945. 225 pp. *P T*

Throughout this impressive work the author proposes theories of interpretation and indicates her tremendous ability to observe and describe movement. She begins with a review of some rare literature on gesture and a chapter on the nature of gesture as a source of information about personality and feelings. There are chapters on the neurophysiology of gesture, developmental factors of gesture (including automatic, stereotyped, and projective gestures of infants and how various emotions are expressed in the infant's movements), and the relation between thinking, imagination, and movement. The greater part of the book is a clinical study of the movements and hand structure of eighty-eight mental patients, including "gesture" case studies of schizophrenic, manic-depressive, and paraphrenic patients. She concludes with a summary of the movement patterns and qualities of various psychiatric syndromes and affective states.

910   **Wolff, Peter H.** "Observations on Newborn Infants." *Psychosomatic Medicine* 21 (1959):110-118. *T*

Continuous observation of four neonates up to five days old, particularly comparing movement and breathing during sleep, drowsiness, and wakefulness, is reported. Six types of activity are defined according to their presumed origins, and a number of specific observations are noted, such as that hand in mouth was used to arrest crying before meals but rarely after, and in what states spontaneous smiles and startles occurred.

911   **Wolff, Peter H.** "The Development of Attention in Young Infants." *Annals of the New York Academy of Science* 188 (1964-65):815-830. *T*

A study of "alert inactivity" in infants (i.e., awake and visually active but motorically quiet), its duration and increase with age, and its relation to attention and contrast with periods of activity and inner stress.

912   **Wolff, Peter H., and Hurwitz, Irving.** "The Choreiform Syndrome." *Developmental Medicine and Child Neurology* 8 (Apr., 1966):160-165. *T*

The incidence of a choreiform twitch as observed in a standardized motor test given to large numbers of children, ages seven to seventeen, some normal, others severely neurotic, delinquent, or blind, was found to be significantly higher in the deviant groups and among boys in every group. The choreiform syndrome, considered a sign of minimal cerebral dysfunction correlating with complications of pregnancy, was observed most often in ten-year-olds, secondly in eleven-year-olds, and to a small degree in twelve-year-olds.

913   **Wolff, Sulammith, and Chess, Stella.** "A Behavioural Study of Schizophrenic Children." *Acta Psychiatrica Scandinavica* 40 (1964):438-466. *T*

Analysis of the histories and observation of fourteen autistic children about five years old. Although the children showed normal motor coordination, a great many of their symptoms involved disturbances in expressive behavior: e.g., staring or avoidance of eye contact, rolling of eyes, rare smiling, rocking and head-banging, and so on.

914   **Wolff, Werner.** "The Experimental Study of Forms of Expression." *Character and Personality* 2 (1933-34):168-176. *P*

Although little is told about the procedures, controls, and quantitative results of the experiments, the findings of this research are fascinating. Subjects tend to recognize their own gait but not their own voice, hands, profile, or mirror-reversed handwriting among a number presented. However, their judgments of their own bodies and expression are different from others' judgments, "go deeper into the personality," and may express subconscious wishes. Further experiments show marked differences between the right and left sides of the face of apes and humans and dramatically different interpretations of the two halves.

915   **Wolff, Werner.** *The Expression of Personality: Experimental Depth Psychology.* New York: Harper & Brothers, 1943. 334 pp. *P T* [New York: Johnson Reprint Corporation, 1971]

A collection of experimental studies of forms of expression and judgment of personality from face, profile, hands, voice, gait, handwriting, and speech. Subjects judged themselves as well as acquaintances and strangers. Evidence of individual style and consistency through different modalities of expression is presented. Studies of judgment of faces indicate high agreement, different interpretations for the right half versus the left half of the face, and dominance of one side or part of the face over another. There are chapters on the judgment of personality from voice, gait, and action and comparison of one's judgments of himself with those of others. Psychological aspects of self-evaluation are explored in depth, including the relation between self-estimation and recall.

916   **Wood, Melusine.** *Some Historical Dances (Twelfth to Nineteenth Century).* London: C. W. Beaumont, 1952. 184 pp. *D*

French, English, and Italian historical dances, "their manner of performance and their place in the social life of the time," are re-created with detailed word descriptions.

917   **Woods, Marcella D.** "An Exploration of Developmental Relationships be-

tween Children's Body Image Boundaries, Estimates of Dimensions of Body Space, and Performance of Selected Gross Motor Tasks." Ph.D. dissertation, Ohio State University, 1966. 206 pp. (Datrix order no. 67-2564)

The "barrier score" from the Holtzman Inkblot Test, estimates and actual mea-surements of body space, and performance in throwing, jumping, and running were correlated for 143 schoolchildren. Age and sex differences accounted for the total variance; however, the author assesses parallel levels of development in body image, motor performance, and cognitive functioning.

918 **Woodworth, Robert S., and Schlosberg, Harold.** *Experimental Psychology.* Rev. ed. New York: Holt, Rinehart, and Winston, 1954. 948 pp. *D G P T*

In this well-written and much-used textbook on problems in psychology, the authors review the main areas of experimental psychology research on movement behavior. They discuss expressive movement primarily in terms of the "activation theory of emotion" and the origin and judgment of facial expression (pp. 107-132); analysis of muscular tension in relation to attention, work efficiency, and thinking (pp. 173-179); and how coordination, fixation, and patterning of eye movements are studied in relation to perception and reading (pp. 492-523).

919 **W.P.A. Project.** *A Bibliography of Dance Origins: A Research into the Anthropological, Ethnological and Religious Origins of the Dance.* Unpublished card index completed in 1936, in the Dance Collection of the New York Public Library, Performing Arts Branch, New York.

An extraordinary collection of annotated references on index cards, filling thirty-eight drawers in the Dance Collection of the New York Public Library. Compiled by a group of performing artists during the Depression era as a W.P.A. project, the collection covers literature up to 1936 concerning dance in cultures all over the world. Extensively indexed by subject and author, a broad range of topics are covered, including psychological aspects, aesthetics, period dances from the Middle Ages on, animal mimicry, sources of dance pictures from Africa to Yugoslavia, children's games and dances, rituals and ceremonial dances from Abyssinia to Zanzibar, and what must be the most extensive reference collection of literature on dance of every culture and country in the world. Praise for this unique collection should be tempered by awareness of its limitations: it exists in only one place, many of the references cited may no longer exist, and it is based on literature available in libraries in the New York area up to 1936 only.

920 **Wulfeck, Wallace H.** "Motor Function in the Mentally Disordered: I, A Comparative Investigation of Motor Function in Psychotics, Psychoneurotics and Normals." *Psychological Record* 4 (1941):271-323. *T*

Schizophrenic, manic-depressive, and psychoneurotic patients and normal controls performed a series of motor tests (e.g., star-tracing, rhythm). Manic patients evidenced higher speed of performance; depressives, psychomotor retardation. Schizophrenics showed profound motor disturbance; psychoneurotics performed as well as normals.

921 **Yang, Martin C.** *A Chinese Village: Taitou, Shantung Province.* New York: Columbia University Press, 1945. 275 pp.

Includes description of a number of social rituals or ceremonies and their significance.

922 **Yarden, Paul E., and Di Scipio, William J.** "Abnormal Movements and Prognosis in Schizophrenia." *American Journal of Psychiatry,* in press. *T*

Eighteen young schizophrenic patients who had abnormal movements described as "choreiform" or "athetoid type," with grimacing, tics, and stereotypies, were compared over three years with schizophrenics who did not have such movement disorders. The patients with movement disorders had significantly earlier onset of illness and much poorer prognoses. The authors postulate that this represents a special clinical entity possibly related to disturbance of Dopamine metabolism and the organic striatal syndromes.

923 **Yates, Aubrey J.** "Abnormalities of Psychomotor Functions." In *Handbook of Abnormal Psychology*, edited by H. J. Eysenck, pp. 32-61. New York: Basic Books, 1961. *G T*

An excellent review of experimental research on the motor-performance differences between normal, neurotic, and psychotic subjects. The studies reviewed involve structured tasks, often with apparatus and sensory stimuli: reaction time, cancellation and tapping tests, tests of instrument-panel control, Luria techniques, mirror drawing, manual dexterity, suggestibility, body sway tests, finger tremor and perseveration measures, measures of performance under fatigue, and tests of fluency, persistence, and strength. A survey of many experimental findings on the motor performance of neurotics, of introverts versus extroverts, and of psychotics is presented together with a valuable bibliography.

924 **Yerkes, Robert M., and Tomilin, Michael I.** "Mother-Infant Relations in Chimpanzees." *Journal of Comparative Psychology* 20 (1935):321-359. *P T*

Summaries of naturalistic observations of five mother-infant pairs from birth to a year old are presented, including behaviors and expressions that occur at different stages; grooming, feeding, and play behavior; how mothers encourage locomotor development, protective behaviors, and intercommunication; and marked individual differences in maternal care and experience.

925 **Yerkes, Robert M., and Yerkes, Ada W.** *The Great Apes: A Study of Anthropoid Life.* New Haven, Conn.. Yale University Press, 1929. 652 pp. *P* [New York: Johnson Reprint Corporation]

A rich source of information about facial and body expressions of emotion, characteristic postures, manual dexterity, locomotion patterns (including note of dance movements), and social behaviors of monkeys, chimpanzees, and gorillas.

926 **Young, Florene M.** "The Incidence of Nervous Habits Observed in College Students." *Journal of Personality* 15 (1946-47):309-320. *T*

The "nervous" hand and face activity of 1,520 college students was observed over six years, using twelve categories adapted from Olson (e.g., twisting hair, biting nails.) Analysis of the results showed an "increase in nervous habits in the war years" (p. 316) and a greater degree of "oral" habits in the girls as compared with the boys.

927 **Young, Paul T.** *Motivation and Emotion: A Survey of the Determinants of Human and Animal Activity.* New York: John Wiley & Sons, 1961. 648 pp. *D G P T*

This experimental psychology text includes a section on muscle tension and its effects on memorizing and practice and a review of theories of emotion that includes analysis of muscle tension, facial expression, and body movement signs of emotion in humans and animals.

928 **Zalk, Mark.** "Movement Behavior in the Theatre: A Study." Unpublished paper, Columbia Teachers College, 1969. 22 pp. *T* (Copies from Dance Notation Bureau, 8 East 12th St., New York)

A study of how differently six actors performed the same movement in a group scene; how an actor moved while acting and not acting; and the differences in the movement of an actress while singing, dancing, and talking. This is a pilot study for a dissertation on the changes in an actor's movement as he develops a character during rehearsal. The author used Laban's effort-shape analysis of movement for the film studies.

929   **Zeligs, Meyer A.** "Acting In: A Contribution to the Meaning of Some Postural Attitudes Observed during Analysis." *Journal of the American Psychoanalytic Association* 5 (1957):685-706.

The author discusses the intrapsychic bases of certain postures and movements observed in psychoanalysis. He describes the female subject's history, course of treatment, and changes in motor behavior.

930   **Zorn, Friedrich A.** *Grammar of the Art of Dancing.* Edited by A. J. Sheafe, Boston: Heintzmann Press, 1905. 302 pp. *D N* [New York: Burt Franklin, Publisher, 1966]

An early German dance notation using stick-figurelike drawings on a staff for leg and arm movements, with thickness of line denoting weight placement and specific symbols for variations such as turns.

931   **Zuk, Gerald H.** "On the Theory and Pathology of Laughter in Psychotherapy." *Psychotherapy: Theory, Research and Practise* 3 (1966):97-101.

Citing patterns of laughter in family therapy sessions, the author proposes that laughter can reflect subgroup alliances, disguise information, mediate relationships, and help convey complex messages at different levels.

SUBJECT INDEX

Indexing by subject has been done from the original works. The numbers listed after each subject refer to the number of the reference in the bibliography. It is possible to find all the sources on a particular topic (e.g., development of facial expression in blind children) by abstracting the reference numbers which are common to each of the relevant subjects (e.g., "blind," "facial expression," "children," and "developmental process"). Boldface type indicates whether the subject is a primary or important one for the reference. It is recommended that the reader first scan the index to see the range of subjects and how they are entered, especially because certain subjects are listed under more general headings.

179

868, 870, 873, 924, 925
Anxiety, 43, 44, 106, **114, 214,** 215,
235, **298, 372, 414,** 467, 525,
531, 576, 590, 595, 596, 599,
622, 658, **706,** 715, 926. *See
also* Psychodiagnostic symptom,
anxiety
Approach-withdrawal, **146,** 174, 524,
568, 733, 734, 773, 774, 886.
See also Attack-defense
Approval, 140, **253, 276, 289, 293,**
519, 551, 584, 617, **651, 652,**
653, 654, 657, 658, 724, 741,
742, 743, 777, 829, 830, 845
Asthma, **214**
Attack-defense, 9, **18, 19, 20,** 21, **50,**
104, 132, **383, 384, 386, 434,**
435, **467, 591, 867, 869, 870.**
See also Approach-withdrawal;
Group interaction, dominance behavior
Attention, 3, 18, 306, 333, **489, 502,**
505, 507, 547, **604, 736,** 762,
911, 918
Attitude, 27, **29, 126, 431, 539,** 547,
651, 652, 653, 654, 655, 656,
657. *See also* Approval
Autogenic therapy, **601,** 775
Automatic writing, **687**

B

Bibliography, **24, 26,** 31, **58, 59, 63,**
117, 152, **173,** 174, **177, 179,**
223, 237, **243, 256, 312, 351,**
389, 408, **423, 431, 434, 440,**
467, 471, 498, 543, 571, 600,
607, 610, 611, **639,** 640, **677,**
682, **701, 704, 773,** 783, 787,
834, **881, 919, 923.** *See also*
Literature review
Bio-energetic analysis, **593, 595,** 776
Blind, 60, **345, 357, 376, 499, 715,**
735, 848, 907, 912
Blinking, **102, 169,** 301, **414, 507,**
590, 715, 734. *See also* Eye
movement
Body attitude. *See* Posture
Body awareness, **5, 6** 128, 131, **152,**
255, **462, 612,** 670, **706, 707,**
718, **776, 784, 786, 821,** 854

Body image, **66,** 147, **152, 153,** 185,
245, 312, 446, **514, 577,** 594,
595, **636, 768, 917**
Body part. *See also* Eye movement;
Facial expression; Posture
arm, **14,** 16, 206, **773**
hand, 10, 13, **17, 34, 75,** 76, **90, 92,**
133, 134, 135, **179,** 213, 218,
219, 221, **224, 311, 314, 356,**
357, 360, 362, **401, 402, 403,**
404, 405, 424, **522, 583, 585,**
586, 608, 614, 615, 635, **641,**
642, 672, 686, 700, 769, **793,**
794, 855, 888, **899, 909, 914**
head, 79, **82, 174, 231, 360,** 390,
502, 505, **518, 551,** 561, **584,**
615, **634, 640, 651, 713, 716,**
741, 742, 743, 761, **818, 864,**
869, 873
leg or foot, **12,** 16, 75, **98, 229, 360,**
484, 635
significance of different parts, 11, 82,
161, **162,** 206, 216, **228, 232,**
250, 261, 262, 269, 406, **409,**
410, 457, **486, 487, 504,** 520,
621, 622, 623, 725, **785, 825,**
901, 915
Body type, **8,** 55, 91, **241,** 254, 388,
464, 529, 570, 789
Brain-damaged. *See* Neurological impairment
Breathing, 114, **152,** 185, **214,** 243,
445, **593,** 594, **595, 596, 615,**
621, **670,** 677, 684, **706,** 707,
725, 726, 746, 775, 784, 786,
825, 854, 910

C

Categories. *See* Definition of behavioral
units
Children, **14, 15, 36, 57,** 65, **66, 72,**
91, 96, 112, **113, 125, 127, 129,**
131, 147, 152, 171, **179, 186,**
189, 222, **226, 239, 240, 241,**
252, 280, **281, 320,** 336, **345,**
349, **353, 354, 355, 357, 358,**
359, 363, **366,** 376, **377,** 386,
421, 425, **439,** 442, 454, **474,**
475, 480, **483,** 490, **491, 493,**
499, 509, 512, 513, 521, **525,**